By the same author

HIGHWAY DESIGN AND CONSTRUCTION

HIGHWAY TRAFFIC ANALYSIS AND DESIGN

HIGHWAY TRAFFIC ANALYSIS AND DESIGN

R. J. Salter

Senior Lecturer in Civil Engineering
University of Bradford

REVISED EDITION

First edition 1974
Revised edition 1976
Reprinted 1978
Reprinted with amendments 1980
First paperback edition with further amendments 1983

Published by
THE MACMILLAN PRESS LTD
London and Basingstoke
Companies and representatives throughout the world

ISBN 0 333 15478 9 (hardcover)
      0 333 36028 1 (paperback)

Printed in Hong Kong

Contents

Preface   vii

PART 1 TRAFFIC ANALYSIS AND PREDICTION

1. Introduction to the transportation planning process   3
2. The transportation study area   8
3. The collection of existing travel data   14
4. The external cordon and screenline surveys   18
5. Other surveys   22
6. Trip generation   25
7. Trip distribution   35
8. Modal split   53
9. Traffic assignment   64
10. The evaluation of transportation proposals   71

PART 2 ANALYSIS AND DESIGN FOR HIGHWAY TRAFFIC

11. The capacity of highways between intersections   95
12. Headway distributions in highway traffic flow   107
13. The relationship between speed, flow and density of a highway traffic
    stream   125
14. The distribution of vehicular speeds in a highway traffic stream   135
15. The macroscopic determination of speed and flow of a highway traffic
    stream   145
16. Intersections with priority control   150
17. Driver reactions at priority intersections   156
18. Delays at priority intersections   163
19. A simulation approach to delay at priority intersections   171
20. Weaving action at intersections   182
21. British and U.S. practice for determining the capacity of higher speed
    weaving sections   189
22. Queueing processes in traffic flow   201
23. New forms of single level intersections   211
24. Grade-separated junctions   219
25. The environmental effects of highway traffic noise   229
26. The environmental effects of highway traffic pollution   245
27. Traffic congestion and restraint   250

CONTENTS

PART 3   TRAFFIC SIGNAL CONTROL

28. Introduction to traffic signals   269
29. Warrants for the use of traffic signals   271
30. Phasing   275
31. Signal aspects and the intergreen period   278
32. Vehicle-actuated signal facilities   280
33. The effect of roadway and environmental factors on the capacity of a
    traffic-signal approach   282
34. The effect of traffic factors on the capacity of a traffic-signal approach   287
35. Determination of the effective green time   291
36. Optimum cycle times for an intersection   294
37. The timing diagram   299
38. Early cut-off and late-start facilities   302
39. The effect of right-turning vehicles combined with straight-ahead and
    left-turning vehicles   306
40. The ultimate capacity of the whole intersection   311
41. The optimisation of signal-approach dimensions   314
42. Optimum signal settings when saturation flow falls during the green
    period   319
43. Delay at signal-controlled intersections   324
44. Determination of the optimum cycle from a consideration of delays on
    the approach   329
45. Average queue lengths at the commencement of the green period   336
46. The co-ordination of traffic signals   342
47. Time and distance diagrams for linked traffic signals   344
48. Platoon dispersion and the linking of traffic signals   349
49. The prediction of the dispersion of traffic platoons downstream of signals   352
50. The delay offset relationship and the linking of signals   354
51. Some area traffic control systems   363

Appendix   371

Index   373

Preface

More than a decade has passed since transport and traffic engineering first became recognised as an academic subject in centres of higher education in the United Kingdom. In this period the interdependence between land use and transport needs has been firmly established and the modelling of the transport system has seen considerable advances.

During this same time it has been realised that highway transport demand in urban areas will have to be balanced against the overall social, economic and environmental costs of movement.

This book includes the fundamental principles of land use, transport planning techniques and the subsequent economic evaluation of highway schemes. It reviews the analytical and practical aspects of highway traffic flow with sections discussing noise generation and pollution and the principles of congestion restraint and road pricing.

Highway intersections are considered in detail. There is a comprehensive treatment of traffic signal control, which ranges from isolated signal-controlled intersections to area-wide signal control.

So that many engineers and planners engaged in transport work, but without the benefits of formal tuition, will find this book useful, each section contains questions by which the reader may test his comprehension of the subject matter by reference to the model answers.

The author would like to express his thanks to those postgraduate students of transportation at the University of Bradford who made valuable comments on the script and examples; also to Professor C. B. Wilby, Chairman of the Schools of Civil and Structural Engineering, for his encouragement for transportation teaching and research.

The author would like to express his appreciation to the following bodies for permission to reproduce their copyright material: Bedfordshire County Council, the Department of the Environment, the Building Research Station, the Eno Foundation, Freeman Fox and Associates, the Greater London Council, Her Majesty's Stationery Office, the Institution of Civil Engineers, the Institution of Highway Engineers, National Research Council, Royal Borough of New Windsor, *Traffic Engineering and Control*, the Transport and Road Research Laboratory, and Wilbur Smith and Associates.

R.J.S.

PART 1
TRAFFIC ANALYSIS AND PREDICTION

1

Introduction to the transportation planning process

Large urban areas have in the past frequently suffered from transportation congestion. It has been recorded that in the first century vehicular traffic, except for chariots and official vehicles, was prohibited from entering Rome during the hours of daylight. While congestion has existed in urban areas the predominantly pedestrian mode of transport prevented the problem from becoming too serious until the new forms of individual transport of the twentieth century began to demand greater highway capacity.

Changes in transport mode frequently produce changes in land-use patterns; for example, the introduction of frequent and rapid rail services in the outer suburbs of London resulted in considerable residential development in the areas adjacent to local stations. More recently the availability of private transport has led to the growth of housing development which cannot economically be served by public transport.

In areas of older development however the time scale for urban renewal is so much slower than that which has been recently experienced for changes in the transport mode that the greatest difficulty is being experienced in accommodating the private motor car. Before the early 1950s it was generally believed that the solution to the transportation problem lay in determining highway traffic volumes and then applying a growth factor to ascertain the future traffic demands.

Many of the early transportation studies carried out in the United States during this period saw the problem as being basically one of providing sufficient highway capacity and were concerned almost exclusively with highway transport.

During the early 1950s however it was realised that there was a fundamental connection between traffic needs and land-use activity. It led to the study of the transportation requirements of differing land uses as the cause of the problem rather than the study of the existing traffic flows. The late 1950s and early 1960s saw the commencement of many land use/transportation surveys in the United Kingdom and the era of transportation planning methodology could have been said to have commenced.

Because the planning of transportation facilities is only one aspect of the overall planning process which affects the quality of life in a developed society, the provision of transport facilities is dependent on the overall economic resources available. It is

3

dependent on the value that is placed on such factors as environmental conditions; for some transport facilities are considered to detract from the quality of the environment and others can be considered to improve the environment. Land use and transport planning are also closely connected because the demand for travel facilities is a function of human land activity and conversely the provision of transport facilities has often stimulated land-use activity.

Because we are living in a society that is changing rapidly, and in which the rate of change appears to be increasing, it is important for some attempt to be made to develop economic, environmental, land use, population and transport planning policies. The fact that planning attempts in all these fields have not met with conspicuous success in the past decade should be taken as an attempt to improve the methodology rather than an indication that short term plans based on expediency or intuition should be employed.

Transportation studies may be carried out to determine the necessity or suitability of a variety of transport systems such as inter-city air-links, a new motorway or a combination of private and public transport modes such as is found in a large urbanised conurbation. The methodology of these surveys will vary in detail—but most transportation surveys that are based on land-use activity tend to be divisible into three major sub-divisions.

(i) The transportation survey, in which an attempt is made to take an inventory of the tripmaking pattern as it exists at the present time, together with details of the travel facilities available and the land-use activities and socio-economic factors that can be considered to influence travel.

(ii) The production of mathematical models, which attempt to explain the relationship between the observed travel pattern and the travel facilities, land-use activities and socio-economic factors obtained by the transportation survey.

(iii) The use of these mathematical models to predict future transportation needs and to evaluate alternative transportation plans. These three stages of the transport planning process are illustrated in figure 1.1, which shows the procedure used to estimate future travel in the Greater London Area.

In the first stage, details of the existing travel pattern together with information on land use and transport facilities are obtained for the area of the study. This area is bounded by an external cordon and so that the origins and destinations of trips within the area can be conveniently described, the study area is divided into traffic zones.

Details of the existing travel pattern are obtained by determining the origins and destinations of journeys, the mode of travel and the purpose of the journey. Most surveys obtain information on journeys that have origins in the survey area by a household interview method, which records details of the tripmaking of survey-area residents. In addition there will be some trips that have origins outside the external cordon and destinations within the cordon and others that have neither origin nor destination within the survey area but pass through the study area. Details of these trips will be obtained by interviewing tripmakers as they cross the cordon. Additional surveys will be necessary to obtain details of commercial vehicle trips originating in the survey area and in some circumstances trips made by means of taxis.

Information on transport facilities will include details of public transport journey times, the frequency of service, walking and waiting times. For the road network, details of traffic flows, journey speeds, the commercial vehicle content and vehicle occupancies are frequently necessary.

As land-use activity is the generator of tripmaking, details of land-use activity are required for each traffic zone. For industrial and commercial land use, floor space and employment statistics are necessary, while for residential areas, the density of development is frequently considered to be of considerable importance. At the same time, socio-economic details of the residents are obtained since many surveys have indicated

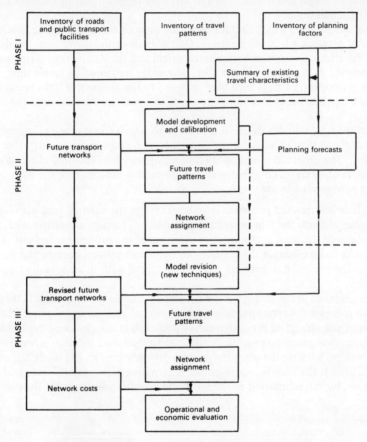

Figure 1.1 Stages of the London Transportation Study

a connection between tripmaking and income, and also between tripmaking and social status.

In the second stage of the transportation process, models are developed which attempt to explain the connection between the tripmaking pattern and land-use activity. This is the most difficult aspect of the transportation planning process in that failure adequately to determine the factors involved in tripmaking will result in incorrect predictions of future transport needs.

While the decision to make a trip is a complex process based on the availability of destinations, the travel facilities, the cost of travel and the journey purpose, it is usual to divide model building into the following interconnected processes which are described in detail in the subsequent sections.

(a) Trip generation, which attempts to determine the connection between tripmaking and land-use factors noted in the planning inventory.

(b) Trip distribution, which determines the pattern of trips between the zones. It is usually postulated that the number of trips between zones is proportional to the size of the zones of origin and destination and inversely proportional to some measure of their spatial separation.

(c) Traffic assignment, which decides on which links of either the public transport or the highway network a trip will be made. On the basis of the travel cost, the decision is made on the alternative routes between the origin and the destination. Trips are usually assigned to the route of least cost, frequently measured by travel time. More recently a proportion of trips have been assigned to the alternative links between an origin and a destination in an attempt to produce a realistic simulation of the real life situation.

(d) Modal split, by which a decision is made as to which travel mode a tripmaker will use. This decision may be made at the trip generation stage or frequently after trip distribution. The observed modal split relationship may be explained in terms of car ownership, relative travel costs between the alternative modes or on the basis of restrained private vehicle use.

After these interrelated processes have been set up, the existing land-use characteristics are inserted into the trip generation procedure. The trips are distributed, assigned and a modal split decision at the appropriate stage of the process carried out. This allows checks to be made on the accuracy of the whole process because the flows assigned to the present day network may be compared with those actually observed on the network.

With confidence in the ability of the developed models established, it is then possible to forecast the travel needs of future land use and transport plans. The evaluation of the effects of these alternative proposals is the object of the third stage of the transportation planning process. Arriving at an optimum solution is however an intuitive process because the planning process can only predict the likely tripmaking which will arise if the plan is implemented. Alternative plans may be evaluated, on a limited basis, by the estimation of the costs and benefits which arise if the plan is carried out.

Problem

During the period 1950–60 an important reappraisal of thinking regarding the estimation of future transportation requirements took place. Select the correct description of this reappraisal from the following and amplify the statement:

(a) It was appreciated that the highway network should be designed to accommodate at least three times as many motor vehicles as are observed on the existing network.

(b) It was appreciated that there was a basic connection between land-use activity and transportation requirements.

(c) It was appreciated that the existing highway network in the centres of large urban areas would require extensive modification.

Solution

(a) While a considerable increase in the number of motor vehicles in use can be expected in the future it is not considered desirable simply to scale up the existing highway network. The demand for transport facilities will increase at differing rates in differing situations and not all of these demands are likely to be met by the provision of enlarged highway facilities.

(b) The realisation that the demand for transport facilities was a function of land use caused an important reappraisal in the estimation of future transport needs. Instead of observing the end product and applying some growth factor to estimate future values it became possible to obtain the traffic requirements of differing land-use plans. It changed transport planning from a vehicle count to a land-use-based operation.

(c) While it was appreciated that the modification of the highway network in the centres of large urban areas would be required in the future it is considered desirable to restrain the use of motor cars in many central areas on environmental grounds.

2

The transportation study area

It is not possible to confine transportation surveys to existing local government bound-
aries because tripmaking and transportation needs are common to a region. Most surveys
are carried out on a regional basis; for example, the West Yorkshire transportation
Study covered an area between Knaresborough and Barnsley, Burnley and Goole. The
London Travel Survey area contains almost 17 per cent of the population of Great
Britain bounded by St. Albans, Slough, Gravesend and Redhill.

The survey area within which travel is to be studied in detail is bounded by an
external cordon. The position of the external cordon is fixed so that as far as possible
all developed areas which influence travel patterns are included together with areas
which are likely to be developed within the forecasting period of the study. Within the
external cordon, trip information is collected in considerable detail and so the cost of
the survey is largely related to the cordon location.

To permit the aggregation of data, the survey area is divided into internal traffic
zones; the remainder of the country external to the cordon is also divided into considerably larger external zones. The Department of the Environment has divided the country
into Standard Regions and this is useful for the external traffic zones.

Zones are selected after consideration of the following factors.

1. They should be compatible with past or projected studies of the region or adjacent
region.
2. They should permit the summarising of land use and tripmaking data.
3. They should allow the convenient assignment of trips, which are assumed to be
generated at the centroids of traffic zones, to the transportation network. There is often
conflict here between zones which are suitable for the highway network and zones
suitable for the public transport network.
4. While greater accuracy is possible with smaller traffic zones they should not be so
numerous as to make subsequent data processing difficult. In the Los Angeles Regional
Transportation Survey there were 2346 'minor' traffic zones which were combined into
274 'major' zones[1]. The London Travel Survey covered an area of 941 square miles and
contained nearly 3 million households. There were 933 zones, 186 districts and 10
sectors[2].
5. The size of the zones will be governed by the size of the survey area and the type of
survey. In a regional survey such as the West Yorkshire Transportation Study the whole
of the Bradford C.B. area was divided into 3 zones and the Leeds C.B. area into 4

8

zones[3]. In contrast for a local study in an urbanised area, where transportation requirements are studied in detail, it is usual to have approximately 1000 households or 3000
people in a zone.

6. Because of the increasing use of data from the Registrar General's Census of Population[4] the zones should be selected after consideration of enumeration, district boundaries
and the grouping of local planning authority data.

7. As data is frequently aggregated by zone, the latter should be areas of predominantly
similar land activity in which similar rates of future growth are anticipated.

When the traffic zones are being determined, it is important to plan them with
thought for the provision of one or more screenlines. A screenline is a line which
divides the area within the external cordon and it is chosen so that the trips crossing it

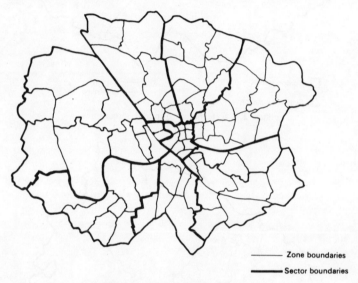

Figure 2.1 Internal traffic zones

at the present day can be easily measured. For this reason a natural barrier to communication should, if possible, be used so as to make it possible to count the traffic movements across the screenline. It should be so positioned that it is unusual for trips to
cross the screenline more than once. To avoid the complex traffic movements which
take place in the central sector a screenline is frequently placed to intercept all the
movements into the central area.

An illustration of a hierarchy of traffic zones is given in figures 2.1, 2.2, 2.3 and 2.4
showing the arrangements used in the Bedford–Kempston Transportation Survey[4].
Figure 2.1 shows the internal zones varying in area from the small intensively developed
Central Area zones to the larger zones in the outer area of Bedford. These internal zones
are grouped into a central area sector and eight radial sectors. The sector numbers 0–8
inclusive are the first digit in the internal zone code numbers.

Within the external cordon, two screenlines were set up for checking purposes and
this is shown in figure 2.2. Also shown are the screenline counting stations and also the
external cordon stations.

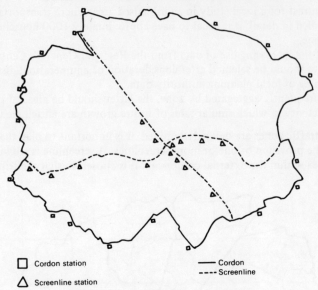

☐ Cordon station   —— Cordon

△ Screenline station   ---- Screenline

Figure 2.2 External cordon and screenlines

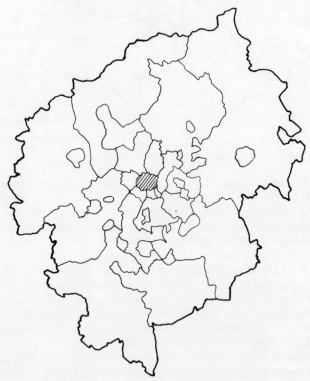

Figure 2.3 External traffic zones (intermediate)

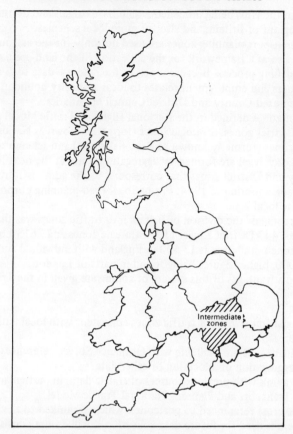

Figure 2.4 External traffic zones (national)

Outside the external cordon but immediately adjacent to the survey area the external zones are comparatively small, generally corresponding to local authority areas or combinations of areas. These zones are shown as intermediate external zones in figure 2.3. The remainder of the country is divided into national external zones as shown in figure 2.4. All the external zones are numbered in accordance with the Department of the Environment standard classification.

Department of Transport National Zoning System

During the mid-1970s the Department of Transport commenced the development of a Regional Highway Traffic Model in order to establish a consistent and rational basis for the prediction of traffic flows on inter-urban highways.

As part of this project a nationally consistent system of zones was defined that was compatible with local authority areas and that could be used for the planning of road networks. A two-stage system was evolved that was suitable both for major studies in which many of the trips being modelled would have origins and destinations in widely spaced regions of Britain, and also for more local schemes.

In land use, transport-planning zones are used not only for the assignment of vehicle movements but also as a framework for the collection of the land-use data that is vital for the modelling process. Because much of the land-use data such as population, household numbers and employment relates to local authority boundaries, the zoning system developed used County and District Council boundaries.

The system of zones defined in the Regional Highway Traffic Model comprised County zones, District zones, Strategic zones (formerly known as National zones), Local (RHTM) zones (formerly known as Regional zones) and scheme zones.

The zones at each level are formed by aggregating zones at the next lower level with the County and District zones that correspond to the administrative area formed after the local reorganisation in 1974. For national road-planning purposes the smallest traffic zone is the local zone.

The final outcome of the division of Britain into traffic zones was that there were 78 County zones, 447 District zones, 1190 Strategic zones and 3613 Local zones. Average Local zone population is 13 500 in England with individual zones ranging from 50 to 50 000, higher figures being found in parts of London.

The expected advantages of this system of zoning are given in the Traffic Appraisal Manual[5].

(a) It is a national framework of traffic zones consistent with local authority administrative areas.
(b) Zone boundary files are available with Ordnance Survey references for plotting at any level of aggregation or allocation of other data.
(c) It provides a link with previous sources of traffic data, in particular the previous Department of Transport and Planning National Traffic Model.
(d) Data can be cross-referenced to postcodes, which are linked to the zone system.
(e) If this system is used for future data collection it will ensure consistency.
(f) Land-use data forecasts are already available to this zoning layout.
(g) The zoning system allows zonal information to be related to an existing or proposed highway network.
(h) A single point location is available which represents the zone area at any zone level.
(i) In the development of the Regional Highway Traffic Model standard computer software has been developed for traffic-modelling purposes using the zone numbering principles.

References

1. Transportation Association of Southern California, Los Angeles Regional Transportation Study, Appendix to the Base Year Report, Los Angeles (1969)
2. London County Council, London Traffic Survey, County Hall, London, Vol. 1 (1964)
3. Traffic Research Corporation, West Yorkshire Transportation Study, Project Report (1967)

4. Bedfordshire County Council, Bedford–Kempston Transportation Survey, Phase 1 (1965)
5. Department of Transport, Traffic Appraisal Manual, London (1981)

3

The collection of existing travel data

Existing tripmaking or travel data may conveniently be classified into four groups according to the origin and the destination of the trip being considered. These four classes of movement are:

(a) trips which have an origin within the external cordon and a destination outside the external cordon;
(b) trips which have both an origin and a destination within the external cordon;
(c) trips which have an origin outside the external cordon and a destination within it;
(d) trips which have neither origin nor destination within the external cordon but which pass through it.

Details of these trips are obtained in differing ways. Trips of type (a) and (b) are usually recorded by means of a home interview survey while trips of type (c) and (d) are normally recorded by means of an origin–destination survey conducted along the external cordon. It is the purpose of this chapter to examine in detail the organisation and analysis of home interview surveys.

The home is the major source of trip generation and details of a considerable proportion of trips generated in an urban area may be obtained by this form of survey.

It is not necessary to obtain details of the trips from every household; instead a sampling procedure may be used. One way in which an unbiased sample may be obtained is by sampling from the electoral roll or from the valuation list, both of which are held by local authorities.

Sample size must depend upon the errors in the data collection process and in the subsequent trip prediction process. Where approximate estimates of travel requirements

Population size	Sample size %
under 50 000	20
50 000–150 000	12·5
150 000–300 000	10
300 000–500 000	6·66
500 000–1 000 000	5
over 1 000 000	4

are necessary, a smaller sample size may be tolerated. The Institute of Traffic Engineers[1] in its Manual of Traffic Engineering Studies recommends sample sizes based on the population of the study area.

Care should be taken to obtain an unbiased sample and if a completed questionnaire cannot be obtained or if it is impossible to deliver a questionnaire then these questionnaires should be omitted from the survey and not delivered at adjacent homes. The trip information obtained must be increased to give all the trips in a zone by the use of a sampling factor

$$\text{sampling factor} = \frac{A - (C \times A/B)}{B - (C + D)}$$

where A = total households in zone
B = number of households selected for sampling in zone
C = number of households vacant or demolished
D = number of households refusing to co-operate

As it is desired to obtain relationships between tripmaking and such variables as land-use characteristics and socio-economic factors, many details other than those relating to the trips must be obtained. The cost of the survey is considerable. Hence great care must be taken in the preparation of these questionnaires both from the point of view of obtaining the required information and also of coding and analysis.

There are alternative approaches to the collection of this data. An interviewer may attempt to contact every member of a selected household individually and details of tripmaking and socio-economic characteristics of the tripmaker are recorded in the presence of the interviewer. Alternatively the interviewer may only collect details of the household characteristics, leaving the questionnaire on tripmaking to be completed at a future date. The interviewer will then collect the questionnaire in the future after

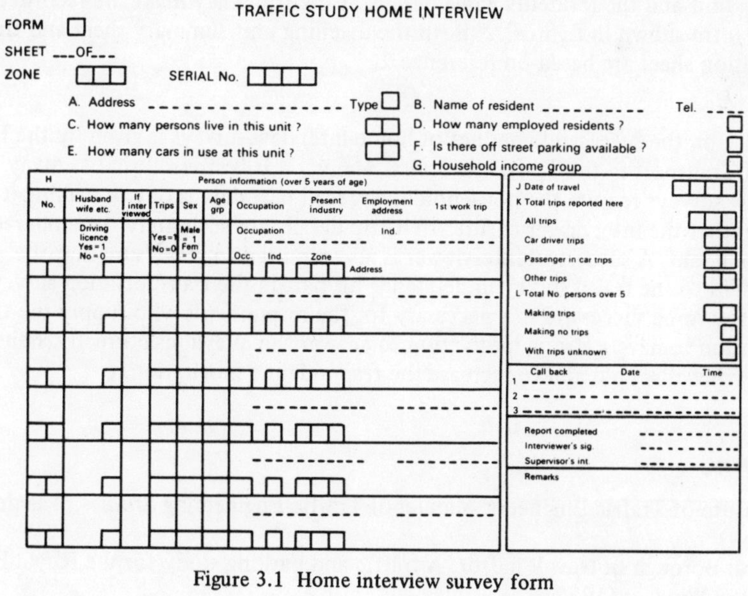
TRAFFIC STUDY HOME INTERVIEW

FORM
SHEET ---OF---
ZONE
SERIAL No.

A. Address ----------------- Type
B. Name of resident ---------- Tel. ----
C. How many persons live in this unit ?
D. How many employed residents ?
E. How many cars in use at this unit ?
F. Is there off street parking available ?
G. Household income group

H					Person information (over 5 years of age)				
No.	Husband wife etc.	If interviewed	Trips	Sex	Age grp	Occupation	Present industry	Employment address	1st work trip
	Driving licence Yes = 1 No = 0	Yes = 1 No = 0		Male = 1 Fem = 0		Occupation / Occ	Ind	Zone / Address	Ind.

J Date of travel
K Total trips reported here
All trips
Car driver trips
Passenger in car trips
Other trips
L Total No. persons over 5
Making trips
Making no trips
With trips unknown

Call back	Date	Time
1		
2		
3		

Report completed
Interviewer's sig.
Supervisor's int.
Remarks

Figure 3.1 Home interview survey form

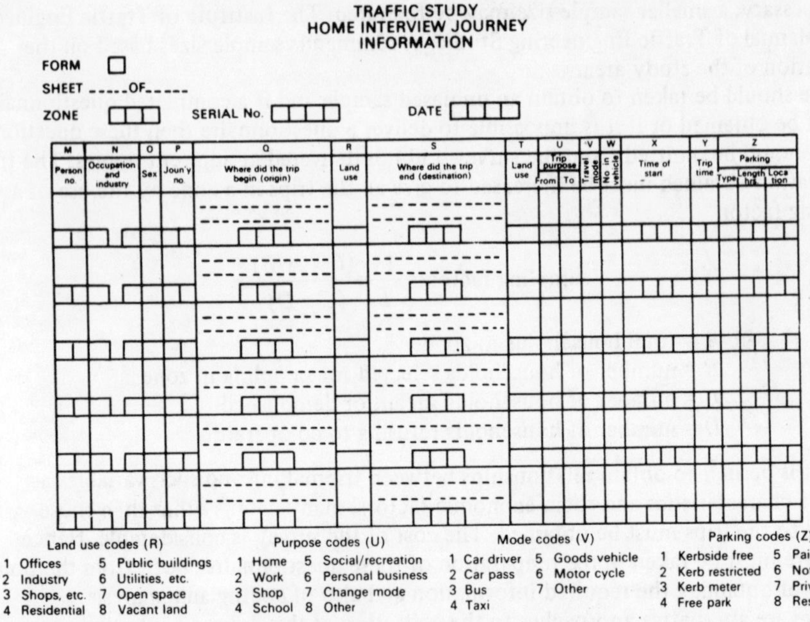

TRAFFIC STUDY
HOME INTERVIEW JOURNEY
INFORMATION

FORM
SHEET .....OF.....
ZONE SERIAL No. DATE

M	N	O	P	Q	R	S	T	U		V	W	X	Y	Z		
Person no.	Occupation and industry	Sex	Joun'y no.	Where did the trip begin (origin)	Land use	Where did the trip end (destination)	Land use	Trip purpose From	Trip purpose To	Travel mode	No. in vehicle	Time of start	Trip time	Parking Type	Parking Length hrs.	Parking Location

Land use codes (R)
1 Offices
2 Industry
3 Shops, etc.
4 Residential
5 Public buildings
6 Utilities, etc.
7 Open space
8 Vacant land

Purpose codes (U)
1 Home
2 Work
3 Shop
4 School
5 Social/recreation
6 Personal business
7 Change mode
8 Other

Mode codes (V)
1 Car driver
2 Car pass
3 Bus
4 Taxi
5 Goods vehicle
6 Motor cycle
7 Other

Parking codes (Z)
1 Kerbside free
2 Kerb restricted
3 Kerb meter
4 Free park
5 Paid park
6 Not parked
7 Private park
8 Residence

Figure 3.2 Home interview journey information

checking them for completeness. While the first method should result in accurate detailing of tripmaking, it is expensive and there is a danger of interviewer prompting. The second method is less expensive in interviewer costs but will tend to be less accurate. Figure 3.1 shows a typical questionnaire used to record the characteristics both of the dwelling unit and the residents while details of the trips they make are recorded on the type of form shown in figure 3.2. Both the dwelling unit summary sheet and the trip information sheet are based on reference 2.

Advice on the successful conduct of home-interview surveys is given by the Department of Transport in the Traffic Appraisal Manual[3]. It is stressed that as many home-interview surveys require the householder to recall past details of trip-making it is essential that the interviewer ensures that trip details are accurately remembered by the respondent. A successful interviewer is also required to make sure that the respondent to the travel survey understands his part in the transportation survey. For any survey to be successful it is necessary for the respondents who supply the trip information to have a strong motivation to answer the questions correctly; hence the interviewer should aim to increase the respondent's movitation.

References

1. Institute of Traffic Engineers, Manual of Traffic Engineering Studies, Washington (1964)
2. Royal Borough of New Windsor, A traffic and parking study for the Royal Borough of New Windsor (1972)

3. Department of Transport, Traffic Appraisal Manual, London (1981)

Problems

In a household-interview survey it is estimated that there are approximately 35 500 households and it is also estimated that not more than 1 in 20 households will, on the survey date, make the particular journey of which the characteristics are to be modelled. If the required accuracy is 2 per cent of the population calculate the required sample size.

Solutions

The Traffic Appraisal Manual[3] gives the following equation for the calculation of sample size, n

$$n = \frac{P(1-P)N^3}{\left(\frac{E}{1.96}\right)^2 (N-1) + P(1-P)N^2}$$

where N is the total number of households within the survey area,
E is the required accuracy expressed as a number of households,
P is the proportion of households with the attribute of interest.

From the above equation

$$n = \frac{0.05 \times 0.95 \times 35\,500^3}{\left(\frac{710}{1.96}\right)^2 (35499) + 0.05 \times 0.95 \times 35\,500^2}$$

$$= 450.$$

4

The external cordon and screenline surveys

The object of the external cordon survey is to obtain information on the trips originating outside the external cordon having origins within the external cordon or passing through the survey area.

For highway trips whether by public or private transport this information is obtained by an origin–destination survey carried out at census points on the roadside. For non-highway public transport trips a survey is carried out while the trips are being made or, alternatively, an examination of ticket records may give some of the required origin–destination information.

Origin–destination surveys at the external cordon may be carried out by interviewing drivers at census points. Alternatively, in certain limited circumstances, the survey may be accomplished by using questionnaire postcards handed to drivers at the census points for later completion and return by post.

The first method, that of direct interview of drivers, is probably one of the most widely used techniques and is commonly used not only to obtain details of trips which cross the external cordon in transportation studies but also for estimating the journeys which would be made on proposed highways.

Direct interviews at roadside census points are often carried out simultaneously at all the census points but a reduction in the number of staff required at any one time, but not in total, and also in the shelters, barriers and warning signs necessary is possible by dealing with different census points and directions of traffic movement at separate times.

Questions asked at the census points will, of course, vary according to the survey, but they should be carefully phrased and without ambiguity. To avoid bias in the results obtained, the questions may sometimes be printed on a card and shown to a driver but it is more usual for the interviewers to be given instructions to ask the exact question required by the traffic engineer in charge of the survey.

A form is usually completed for each interview, the enumerator noting the time, the census point and the direction being sampled, the class of vehicles, the origin and destination of the journey and any intermediate stops. If the economic benefits of the proposed highway are to be assessed then additional questions may be asked to determine the purpose of the journey and the number of occupants of the vehicles. A typical interview sheet used for recording details of vehicle trips is given in figure 4.1.

Under normal traffic conditions it is of course not possible to interview the driver of each vehicle without causing undue delay or employing an excessive number of interviewers. For this reason only a proportion of the vehicles travelling past a census point are stopped for interview, a process known as sampling.

The selection of the vehicles for the sample must be free from bias and the sampling technique developed by the Transport and Road Research Laboratory has been widely used[1].

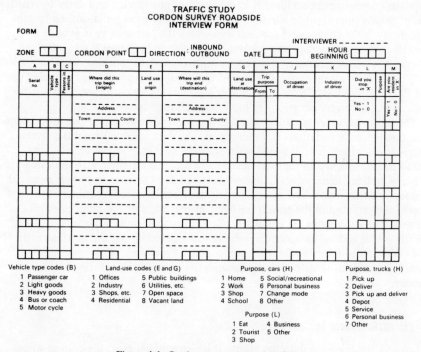

Figure 4.1 Cordon survey interview form

In this method a police officer, upon receiving a signal from either of the two interviewers that he is ready to commence the next interview, directs the first vehicle that arrives at the census point into the interview lane. All other traffic is allowed to pass the census point without being halted, a volume count of the number of vehicles passing the point being carried out by the enumerator.

If the first vehicle to arrive at the census point after an interviewer has indicated that he is ready to carry out a further interview is halted, then the sampling technique is free from bias. It has been found, however, that on derestricted lengths of highway the first vehicle may be travelling at a high speed and the police officer directing the traffic may halt a slower vehicle which is following. Should this tendency not be eliminated by slowing down the whole of the traffic stream then the sample is biased in favour of slower vehicles.

As no attempt is made to ensure that the number of vehicles halted is a constant proportion of the total number of vehicles passing the census point, a sampling factor is introduced for each class of vehicle which varies according to the hour. The sampling

factor for a certain hour is simply the number of vehicles of a particular class passing
the census point during that hour divided by the number of vehicles of the particular
class included in the sample.

The growth of traffic that has occurred since the publication of reference 1 has
led to the issue of detailed instructions for the conduct of traffic surveys by roadside
interview[2].

General instructions are given on the location of census points with particular
emphasis on minimising accident risk, carriageway obstruction and delay to vehicles.
It is obviously important to site the census point away from junctions, on lengths
of highway without excessive gradient and where the carriageway is as wide as
possible. For single carriageway roads it is desirable to incorporate a bypass lane
otherwise it will be necessary to operate stop/go or shuttle working. There are fewer
difficulties when a census point is located on a dual carriageway road provided speeds
and flows are not high. In the British Isles interviewing on motorways has not been
allowed and only very occasionally has it been allowed on slip roads.

The information requested at roadside interview surveys carried out in accordance
with these recommendations is standardised. The data collected from the drivers
sampled comprises: the full postal address of the next stop of the vehicles; the
reason for the journey to the next destination which is classified into the categories
of home, holiday home, work, employer's business, education, shopping, personal
business, visiting friends and recreation/leisure. Details are also required of the last
stop of the vehicle and the reason for being at that address; the same categories as
are used for destinations are employed. Car, motorcycle and van drivers are asked
the full postal address at which the vehicle was garaged during the previous night.

In addition to interviewers it is also necessary for enumerators to undertake
classified counts of vehicles passing through the census station during half-hour
periods.

Registration number surveys

If information is required on the points at which vehicles cross the cordon when
entering or leaving the study area but there are difficulties in stopping vehicles, then
a registration number survey can be carried out. Registration numbers and times
are noted as vehicles cross the cordon, and subsequently these numbers are matched
up to give an indication of vehicle paths through the area.

In order to reduce the amount of writing and hence improve the accuracy of the
survey it is usual to write only a portion of the registration number, normally the
right-hand numerical section together with the year letter. When a registration number
cannot be seen then its passage is noted with a tick. Classification of vehicles is made
by placing an additional identifying letter after the registration number.

At heavily trafficked situations it may not be possible for an observer to record
all the registration numbers. In these circumstances tape recorders are a useful means
of quickly recording details, a time base being conveniently incorporated into the
record. Matching of the registration numbers can be laborious, especially if many
vehicle registration numbers have been missed or wrongly noted.

Using postcards

The use of prepaid postcards which are handed to drivers as they pass the census points is an inexpensive method of obtaining journey information. Delay to vehicles is small and a considerable amount of information may be obtained from the questions on the postcard[2]. A great deal of advance publicity however is required if a reasonable proportion of postcards are to be returned, and often the number of cards returned has been disappointing. There is also the danger that certain types of vehicle drivers are more likely to complete and return the postcards than others, so giving a bias to the results obtained.

Details of the trips crossing the screenlines within the external cordon are necessary to allow a check to be made on the information obtained by the home-interview, external cordon and goods movement surveys. The total existing trip information is assigned to the present day or base year transport networks and if a discrepancy is found then the survey information is adjusted. Screenline checks are also useful to compare sector-to-sector movements as a check on the trip generation, attraction and distribution models developed from the present day pattern of movement.

References

1. Transport Road Research Laboratory, *Research on Road Traffic,* Chapter 4, HMSO, London (1965)
2. Department of Transport, *Traffic Surveys by Roadside Interview,* Advice Note TA/11/81, London (1981)
3. P.A. Caxton, Improved techniques for registration number traffic surveys, *Traff. Engng. Control* (June 1967)

5

Other surveys

Commercial vehicle surveys

This survey is designed to measure the trips made by commercial vehicles within the internal area. The data is obtained by issuing drivers with forms on which they record each trip, together with trip origin, destination, purpose and parking details. The size of the sample depends on the size of the survey area and also on the variability of goods-use activity, ranging from 100 per cent for a small town to about 25 per cent for a large area. The population may be obtained from the land-use survey, which will indicate addresses at which commercial vehicles may be kept or alternatively excise licence records may be consulted. A typical questionnaire is reproduced in figure 5.1.

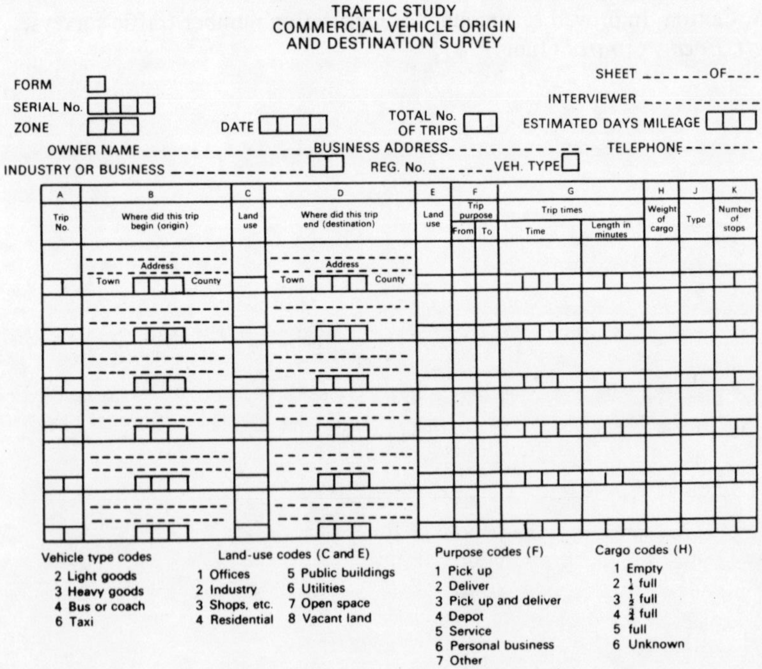
TRAFFIC STUDY
COMMERCIAL VEHICLE ORIGIN
AND DESTINATION SURVEY

SHEET _______OF____

FORM
SERIAL No.
ZONE
DATE
TOTAL No. OF TRIPS
INTERVIEWER ___________
ESTIMATED DAYS MILEAGE
OWNER NAME ________________ BUSINESS ADDRESS ____________ TELEPHONE ________
INDUSTRY OR BUSINESS _____________ REG. No. _______ VEH. TYPE

A	B	C	D	E	F		G		H	J	K
Trip No.	Where did this trip begin (origin)	Land use	Where did this trip end (destination)	Land use	Trip purpose		Trip times		Weight of cargo	Type	Number of stops
					From	To	Time	Length in minutes			
	Address Town County		Address Town County								

Vehicle type codes
2 Light goods
3 Heavy goods
4 Bus or coach
6 Taxi

Land-use codes (C and E)
1 Offices
2 Industry
3 Shops, etc.
4 Residential
5 Public buildings
6 Utilities
7 Open space
8 Vacant land

Purpose codes (F)
1 Pick up
2 Deliver
3 Pick up and deliver
4 Depot
5 Service
6 Personal business
7 Other

Cargo codes (H)
1 Empty
2 ¼ full
3 ½ full
4 ¾ full
5 full
6 Unknown

Figure 5.1 Commercial vehicle survey form

Bus passenger surveys

Bus trips made by residents are obtained from the home interview survey; trips made by public transport by non-residents are obtained by a bus passenger survey carried out at the external cordon. Duplication of trips made by residents within the survey area should be checked.

Questionnaires are usually distributed to bus passengers as the bus enters the survey area at the external cordon. The completed questionnaires are collected from the bus or passengers are requested to return the completed forms by post. The questionnaire used in the Leicester Traffic Survey[1] is given in figure 5.2 as an illustration.

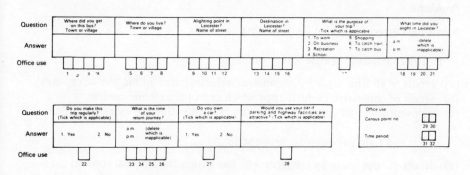

Figure 5.2 A typical bus passenger questionnaire

In the larger urban areas there will also be a considerable number of non-home-based trips within the central area by tripmakers who are not resident within the external cordon. As these trips will not be recorded by the home interview survey it is necessary to either conduct additional surveys within the external cordon or interview a sample of tripmakers leaving the survey area.

Train passenger surveys

Questionnaires are frequently distributed to train passengers as they board trains bound for the survey area. Questionnaires are collected on the train or passengers are requested to return them by post. When postal return is used the information obtained should be checked wherever possible when the return is less than 70 per cent.

The information requested on a train passenger survey will vary according to the survey type. It will include questions requesting information on:

(a) the address of the traveller;
(b) the mode of travel from the home to the station;
(c) the scheduled departure time of the train;
(d) the mode of travel from the destination station to the workplace;
(e) the address of the destination;
(f) the journey purpose;
(g) details of vehicle ownership.

Taxi surveys

If required a survey is carried out as for a commercial vehicle survey. Normally taxi trips form only a small proportion of tripmaking in most cities and a taxi survey is not normally necessary unless the number of trips by this mode of transport are considerable

Reference

1. J. M. Harwood and V. Miller, *Urban Traffic Planning*, Printerhall, London (1964)

Problems

Select the correct answer to the following questions.

    1. The commercial vehicle survey obtains:

(a) details of all trips made by commercial vehicles based within the survey area;
(b) details of commercial vehicle trips which cross the external cordon;
(c) an inventory of all commercial vehicles garaged within the external cordon.

    2. Bus and train passenger surveys obtain:

(a) details of trips made by residents who live within the external cordon;
(b) details of trips made into and through the survey area by tripmakers resident outside the external cordon;
(c) details of journey times by public transport within the survey area.

Solutions

The correct answers to the above questions are:

    1. The commercial vehicle survey obtains:

(a) details of all trips made by commercial vehicles based within the survey area.

    2. Bus and train passenger surveys obtain:

(b) details of trips made into and through the survey area by tripmakers resident outside the external cordon.

6

Trip generation

Once the transportation survey has collected all the details of the existing tripmaking pattern and the socio-economic, land-use and transportation-system characteristics of the survey area, the second stage in the transportation planning process is the development of relationships between the total number of trip origins and destinations in a zone and the zonal characteristics. It is assumed that these relationships will be true in the future and so, if land-use and socio-economic factors can be predicted, future trips can be estimated for any proposed transport system.

At this stage some basic definitions have to be made. A trip is a one-way person movement by one or more modes of travel and each trip will have an origin and a destination. Most surveys divide trips into home-based and non-home-based trips. All trips which have one end at the home are said to be generated by the home and the

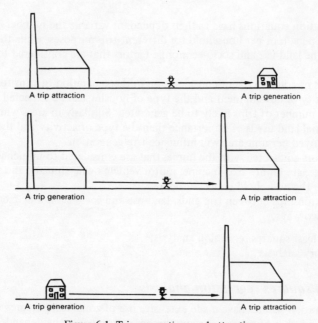

Figure 6.1 Trip generations and attractions

other end of the trip is said to be attracted to the zone in which it commences or terminates. For non-home-based trips the zone of origin is said to generate the trip and the zone of destination is said to attract the trip. Figure 6.1 illustrates the differences between generation and attraction.

The form of the relationship connecting trip generation and land use and socio-economic factors depends upon the basic framework of the transportation process. If the study is designed to use trip end modal-split models, which allocate trips to differing modes of transport before the distribution of trips between the traffic zones, then trip generation models are designed to predict person movement by differing modes of travel. On the other hand when trips are distributed between the zones before the decision is made regarding the mode of travel, then a trip interchange model is required, which predicts person movement in terms of total person movement by all modes of travel.

Because trips made for different purposes have different distribution and modal split characteristics, trip generation equations are generally stratified into trips for different purposes whether trip end or trip interchange models are being developed.

As the home is regarded as being the generator of the major proportion of trips, a considerable number of trips have the home end of the trip as a part of the description; that is

home to work trips
home to recreation trips

and where the land use at the destination does not fall into any of the stated categories then the description of

home to other trips

is used.

Trip generation equations have as their dependent variable the number of trips generated per person or per household for different trip purposes, while the independent variables are the land-use and socio-economic factors that are considered to affect trip-making.

Land use is of course a major consideration in the generation of trips for in residential areas the density of development and the type of housing can be expected to be major factors in the number of trips likely to be generated. Similarly in areas where commercial or industrial land use is of importance then the type of activity and the number of workers employed per unit area will influence trip generation.

Other factors connected with the home, that are considered to influence the trip generation rate, are family size, income, motor vehicle ownership and the social status of the head of the household being considered.

These relationships between trip ends, land-use and socio-economic factors may be obtained in two ways:

(1) by zonal least squares regression analysis;
(2) by category analysis.

Zonal least squares regression analysis

Past transportation studies have made extensive use of zonal least squares regression. For each of a number of zones a certain number of trip ends—the dependent variable—

are observed and each zone has certain measurable characteristics to which this trip generation rate may be related. These characteristics X_1, X_2 etc., are referred to as the independent variables and are the land-use and socio-economic factors which have been previously referred to.

The equation obtained by least squares analysis is of the general form

$$Y = b_0 + b_1 \times X_1 + b_2 \times X_2 + \cdots + b_n \times X_n$$

where b_0 is the intercept term or constant,

$$b_0, b_1 \ldots b_n \text{ are obtained by regression analysis,}$$

$$X_1, X_2 \ldots X_n \text{ are the independent variables.}$$

In developing regression equations it is assumed that:

1. All the independent variables are independent of each other.
2. All the independent variables are normally distributed; if the variable has a skew distribution often a log transformation is used.
3. The independent variables are continuous.

It is not usually possible for the transportation planner to conform to these requirements and regression analysis has been subjected to considerable criticism.

Nevertheless regression analysis is a powerful tool when used in conjunction with a computer for handling considerable volumes of data.

It is important however to recognise that the regression process contains the likelihood of the future values of the dependent variable Y being in error when future values of the independent variables X_1, X_2, etc., are substituted into the equation. The likely sources of error may be stated to be:

(a) errors in the determination of the existing values of the independent variables owing to inaccuracy or bias in the transportation survey;
(b) errors in the determination of the existing values of the dependent variables, also as a result of inaccuracy or bias in the transportation survey. This may be detected and corrected by adequate screenline checks.
(c) the assumption that the regression of the dependent variable on the independent variables is linear, a matter of some importance when future values of the independent variables are outside the range of observed values;
(d) errors in the regression obtained owing to the scatter of the individual values and the inadequacy of the data;
(e) difficulties in the prediction of future values of the independent variables, for the future value of the dependent variable will only be as good as the future estimates of the independent variables;
(f) future values of the independent variable will be scattered as are the present values;
(g) the true regression equation may vary with time because factors that exert an influence on tripmaking in the future are not included in the present-day regression equation.

Most computer programs introduce or delete the dependent variables in a stepwise manner. Only variables that have a significant effect on the prediction of the dependent variables are included in the regression analysis.

It is usual to compute the following statistical values to test the goodness of fit of the regression equation.

1. Simple correlation coefficient r which is computed for two variables and measures the association between them. As r varies from -1 to $+1$ it indicates the correlation between the variables. A value approaching ± 1 indicates good correlation.
2. Multiple correlation coefficient R which measures the goodness of fit between the regression estimates and the observed data. $100R^2$ give the percentage of variation explained by the regression.

Transportation studies have produced a considerable number of regression equations and their variety is often confusing. This is partly owing to variations in the form of the independent variables which have been used and also to variations in tripmaking.

A typical equation obtained in the Leicester transportation study[1] was

$$Y_8 = 0.0649X_1 - 0.0034X_3 + 0.0066X_4 + 0.9489Y_1$$

where Y_8 = total trips per household where the head of the household is a junior non-
         manual worker/24 h
    X_1 = family size
    X_3 = residential density
    X_4 = total family income
    Y_1 = cars/household

If all the factors influencing the pattern of movement are correctly identified then it might be expected that trip generation equations obtained in one survey would be applicable to other surveys. Attempts to establish similarity between relationships observed in different surveys have not however met with any great success because of variations in both the dependent and independent variables chosen.

Category analysis

Difficulties with the use of regression equations for the study of trip generation have led to considerable support being given to the use of disaggregate models, that is, models based on the household or the person. Making use of the unexpanded sample data, these models make no reference to zone boundaries and allow considerable flexibility in the selection of alternative zone systems when future trip ends are being predicted.

This approach has become known as category analysis and has been largely developed by Wootton Pick and Gill[2,3] and has been applied to a considerable number of transport studies. Category analysis uses the household as the fundamental unit of the trip generation process and assumes that the journeys it generates depend on household characteristics and location relative to workplace, shopping and other facilities. Trip generation is measured as the average number of one-way trips generated by a household on an average weekday.

Household characteristics that are readily measured and appear to account for variation in generation both at the present time and in the future are disposable income, car ownership, family structure and size. Location characteristics have proved more difficult to isolate and the characteristic that has found the greatest application is public transport accessibility.

Wootton and Pick[2] classified households according to:

1. Cars owned—(1) None
   (2) 1
   (3) More than 1
2. Income —Originally 6 classes were proposed ranging from less than £500 to £2500 or more. These classes were modified in the West Midland Study to give an income class up to £10 000. There is usually a difficulty in obtaining sufficient observations in higher income groups and yet it is in these ranges that future prediction is important.
3. Family structure

	Adults employed	Adults not employed
(1)	None	1
(2)	None	More than 1
(3)	1	0, 1
(4)	1	More than 1
(5)	More than 1	0, 1
(6)	More than 1	More than 1

These categories produced 108 household classes and associated with each class is a trip rate. It was proposed that trips be classified by mode of travel and trip purpose. These are:

Modes 1. Drivers of cars or motor cycles
2. Public transport passengers
3. Other passengers (mostly car passengers)
Purpose 1. Work   2. Business   3. Education
4. Shopping   5. Social   6. Non-home based

where there are thus 6 x 3 = 18 mode and purpose trip combinations.

The basic assumption is that trip rates are stable over time and that the future behaviour of a household can be described by the category into which it falls. The number of households in each category can be estimated by the fitting of mathematical distributions to the observed values of income, car ownership and family structure.

Income distribution

Pick and Gill have shown that income distribution may be represented by a continuous probability density function $Q(x)$ such that the number of households having income $x, a < x < b$, is given by

$$N \int_a^b Q(x) \, dx$$

where N is the number of households in the zone.

The following distribution has been used for $Q(x)$

$$Q(x) = \frac{\alpha^{n+1}}{n!} x^n e^{-\alpha x}$$

where $\alpha = \bar{x}/s^2$,

     $n = (\bar{x}^2/s^2) - 1$,

     $\bar{x}$ is the mean income,

     s^2 is the standard deviation of income.

Future incomes were projected on the basis of

$$x^1 = x(1 + g)^y$$

where g is the annual growth rate of income relative to the cost of living,

     y is the number of years projected,

     x^1 is the future income,

     x is the present income.

Car ownership distribution

A conditional probability function $P(n/x)$ was derived from the West Midlands Transportation Study to give the probability of a household owning n cars if its income relative to the price of cars is x.

     Then

$$P(0/x) = Ke^{-\beta x}$$

$$P(1/x) = Ce^{-\beta x} \times (\beta x)^n$$

where K and C are constants and β varies with bus accessibility.

     Variations of $P(0/x)$ and $P(1/x)$ within a study area as given by Pick and Gill are shown in figure 6.2. These variations have been explained by the following factors, but more detailed information can be obtained from reference 3.

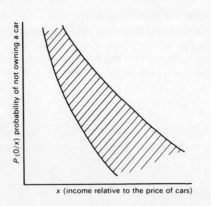 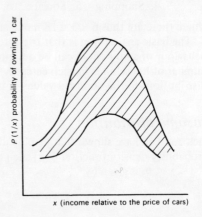

Figure 6.2 Probability of car ownership (adapted from ref. 3)

(a) An increase in residential density produces a higher $P(0/x)$ curve and a lower $P(1/x)$ curve. It has been found that the effects of high and low density are well defined but there is some variation in the medium density ranges

(b) An increase of the public transport accessibility index of a traffic district causes an increase in the value of $P(0/x)$. It is defined as

$$\sum_j \sqrt{b_j}/\sqrt{a_i}$$

where a_i = area of district i (sq. miles)

$\quad\quad b$ = number of buses on route j passing through the district per unit time.

This index has a value of 0 in a rural area without bus services and a value of approximately 60 in central London.

(c) The cost of housing has been bound to influence car ownership: the greater the housing cost the higher the $P(0/x)$ curve.

(d) Lack of garage space inhibits multi-car owners and influences both the $P(0/x)$ and $P(1/x)$ curve.

(e) Because older household members are likely to have more money available for car purchase and younger household members are likely to be moped and motor cycle owners, age structure influences car ownership curves.

(f) Spatial relationships affect car ownership for if activities are closely related to housing then there is less necessity for the purchase of vehicles.

Family structure distribution

As previously stated households are split into 6 categories and the number of households in each category may be estimated using the following argument.

Let the probability of a household having n members be $Q(n)$.

Let $p(n)$ be the probability that a member of an n member household is employed. The probability that r members of an n member household are employed is

$$\frac{n!}{r!(n-r)!} p(n)^r (1 - p(n))^{n-r}$$

assuming a binomial distribution. The probability that a household has n members of whom r are employed is then

$$P(n, r) = \frac{n!\, Q(n)}{r!\, (n-r)!}\, p(n)^r (1 - p(n))^{n-r}$$

where $Q(n) = \dfrac{e^{-x} x^{n-1}}{(n-1)!}$ $\quad\quad n = 1, 2, ...$

$\quad\quad x$ = average family size $- 1$

$\quad p(n) = \dfrac{\text{employed residents}}{\text{total households}}.$

Category analysis applied to the Los Angeles Regional Transportation Study

The LARTS transportation model[4] analyses five different types of car-driver trips (the dominant mode). Three of the trip types are home based and account for 67.2 per cent of total trips. Since all travel was associated with a home base for purposes of computation, the non-home-based trips were distributed at a later stage to non-home-based zones mainly in direct proportion to each zone's share of total sector non-home-based activity, as determined by the amount of retail employment and population in the traffic zone.

The car drive trip generation rates (weekday passengers vehicle trips per vehicle) are given in table 6.1.

TABLE 6.1 Car-drive trip generation rates (from reference 4)

Dwelling Unit Category	Trip Type 1 Home-other	2 Other/Other	3 Work/Other	4 Home/Work	5 Home/ Shopping
ONE VEHICLE					
Single family	*0·98–1·55	0·50–1·12	0·25–0·53	0·46–1·06	0·85
Multi family	0·72–1·24	0·71	0·45–0·52	0·88–1·05	0.44
TWO OR MORE VEHICLES					
Single family	0·90–1·12	0·81	0·33–0·43	0·17–0·91	0·52
Multi family	0·50–1·12	0·67	0·62–0·31	0·17–0·91	0·37
NO VEHICLE†					
Single family	0·49	0·28	0·08	0·21	0·18
Multi family	0·02	0·03	0·04	0·06	0·03

* Range of trips depends on percentage of adults age 60 or over; 0.98 if 100 per cent over 60; 1.55 if 0 per cent over 60.
† Rate applied to number of dwelling units for this category.

Category analysis applied to the London Transportation Study

From an analysis of household characteristics obtained during the London Transportation Study[5] it was found that the following factors affected trip generation:

1. Car ownership.
2. Employed persons.
3. Income.
4. Residential density.

TABLE 6.2 24 hour average weekday generation rates for residents of households with one employed resident in areas of medium rail and bus accessibility (from reference 6)

Income Class	No car	1-car	2 cars or more
Low	2·06 – +1·99 –4·05		
	+1·01	+1·24	
Medium	3·07 – +2·22 –5·29 – +2·32 – 7·61		
High	+1·34	+3·34	
		6·63 – + 4·32 – 10·95	

Each factor was divided into three categories giving a total of $3 \times 3 \times 3 \times 3 = 81$ household categories. To estimate future generation in the survey the planning parameters for each district were used to estimate the number of households in each category and these were then multiplied by the appropriate generation rates. Table 6.2 gives values for residents of households with one employed resident in areas of medium rail and bus accessibility.

Advantages and disadvantages of the category analysis technique

A great advantage of the category analysis technique is that it is possible to estimate household categories from the Registrar General's Census Data using known relationships. Trip generation rates obtained from other surveys are then used subject to a small survey check on the accuracy of the rates. When the cost of large scale home interview surveys is considered, the advantage of this technique can be appreciated.

The computational techniques are also simpler than those required for zonal least squares regression. The use of disaggregate data may also be expected to simulate human behaviour more realistically than zonal values.

A disadvantage of the category analysis technique however is that it is assumed that income and car ownership will increase in the future. The categories with high incomes and car ownership are however the ones which are least represented in the base year data. Moreover they are the categories which are most likely to be used for future estimates of trip generation.

References

1. J. M. Harwood and V. Miller. *Urban Traffic Planning*. Printerhall, London (1964)
2. H. J. Wootton and G. W. Pick. Travel estimates from census data. *Traff. Engng Control*, **9** (1967), 142–5
3. G. W. Pick and J. Gill. New developments in category analysis. PRTC Symposium, London (1970)
4. Transportation Association of Southern California. LARTS 1980. Progress Report, Los Angeles
5. London County Council. London Traffic Survey, 1 (1964)
6. Greater London Council. Movement in London. County Hall, London (1969)

Problems

Are the following statements true or false?

(a) A trip with an origin at the workplace and a destination at the home is said to be generated by the home.

(b) A journey from work to home made by walking to the bus, travelling by bus to the station and completing the journey by train is regarded as three trips.

(c) A trip end modal-split generation model predicts the trips generated by a traffic zone regardless of the mode of travel.

(d) In a trip generation equation the independent variables usually describe land-use and socio-economic factors.

(e) There would be no objection to the use of total household income and number of employed members in the household as independent variables in a trip generation equation.

(f) The use of trip generation equations to predict future trips depends on the ability to estimate future values of the independent variables.

(g) Category analysis uses disaggregate survey data while regression analysis employs zonal aggregate survey data.

(h) The examination of survey data by the use of category analysis techniques shows that the trip generation rate decreases with car ownership and increases with income class.

Solutions

Statement (a) is correct since all work trips with an origin or a destination at the home are said to be generated at the home.

Statement (b) is incorrect since a trip is a single journey between an origin and a destination.

Statement (c) is incorrect since a trip end modal-split model predicts trips classified by modal type; that is, modal split is carried out before trip distribution takes place.

Statement (d) is correct because trip generation equations have as their dependent variable the number of trips generated while the independent variables are the land-use and socio-economic factors which affect the generation of trips.

Statement (e) is incorrect in that the independent variables should be independent of each other while there is likely to be a strong correlation between the number of employed members of a household and the total household income.

Statement (f) is correct because future trips are estimated by assuming that the same correlation as exists today between tripmaking and land use and socio-economic factors will exist in the future. Future trips are then estimated by substituting future land-use and other factors into the trip generation equations.

Statement (g) is correct in that category analysis attempts to predict tripmaking at the household level while least squares regression analysis derives equations from aggregated zonal data.

Statement (h) is incorrect since the analysis of the tripmaking habits of households shows that the number of trips made by households increases as the number of cars owned by the household increases and as household income increases.

7

Trip distribution

Trip distribution is another of the major aspects of the transportation simulation process and although generation, distribution and assignment are often discussed separately, it is important to realise that if human behaviour is to be effectively simulated then these three processes must be conceived as an interrelated whole.

In trip distribution, two known sets of trip ends are connected together, without specifying the actual route and sometimes without reference to travel mode, to form a trip matrix between known origins and destinations.

There are two basic methods by which this may be achieved:

1. Growth factor methods, which may be subdivided into the

    (a) constant factor method;
    (b) average factor method;
    (c) Fratar method;
    (d) Furness method.

2. Synthetic methods using gravity type models or opportunity models.

Trip distribution using growth factors

Growth factor methods assume that in the future the tripmaking pattern will remain substantially the same as today but that the volume of trips will increase according to the growth of the generating and attracting zones. These methods are simpler than synthetic methods and for small towns where considerable changes in land-use and external factors are not expected, they have often been considered adequate.

(a) *The Constant Factor Method* assumes that all zones will increase in a uniform manner and that the existing traffic pattern will be the same for the future when growth is taken into account. This was the earliest method to be used, the basic assumption being that the growth which is expected to take place in the survey area will have an equal effect on all the trips in the area. The relationship between present and future trips can be expressed by

$$t'_{ij} = t_{ij} \times E$$

35

where t'_{ij} is the future number of trips between zone i and zone j. t_{ij} is the present number of trips between zone i and zone j. E is the constant factor derived by dividing the future number of trip ends expected in the survey area by the existing number of trip ends.

This method suffers from the disadvantages that it will tend to overestimate the trips between densely developed zones, which probably have little development potential, and underestimate the future trips between underdeveloped zones, which are likely to be extensively developed in the future. It will also fail to make provision for zones which are at present undeveloped and which may generate a considerable number of trips in the future.

(b) *The Average Factor Method* attempts to take into account the varying rates of growth of tripmaking which can be expected in the differing zones of a survey area.

The average growth factor used is that which refers to the origin end and the destination end of the trip and is obtained for each zone as in the constant factor method. Expressed mathematically, this can be stated to be

$$t'_{ij} = t_{ij} \frac{(E_i + E_j)}{2}$$

where $E_i = \dfrac{P_i}{p_i}$ and $E_j = \dfrac{A_j}{a_j}$

t'_{ij} = future flow ab,

t_{ij} = present flow ab,

P_i = future production of zone i,

p_i = present production of zone i,

A_j = future attraction of zone j,

a_j = present attraction of zone j

At the completion of the process attractions and productions will not agree with the future estimates and the procedure must be iterated using as new values for E_i and E_j the factors P_i/p'_i and A_j/a'_j where p'_i and a'_j are the total productions and attractions of zones i and j respectively, obtained from the first distribution of trips. The process is iterated using successive values of p'_i and a'_j until the growth factor approaches unity and the successive values of t'_{ij} and t_{ij} are within 1 to 5 per cent depending upon the accuracy required in the trip distribution.

The average factor method suffers from many of the disadvantages of the constant factor method, and in addition if a large number of iterations are required then the accuracy of the resulting trip matrix may be questioned.

(c) *The Fratar Method*[1] This method was introduced by T. J. Fratar to overcome some of the disadvantages of the constant factor and average factor methods. The Fratar method makes the assumptions that the existing trips t_{ij} will increase in proportion to

E_i and also in proportion to E_j. The multiplication of the existing flow by two growth factors will result in the future trips originating in zone i being greater than the future forecasts and so a normalising expression is introduced which is the sum of all the existing trips out of zone i divided by the sum of all the existing trips out of zone i multiplied by the growth factor at the destination end of the trip. This may be expressed as

$$t'_{ij} = t_{ij} \times \frac{P_i}{p_i} \times \frac{A_j}{a_j} \times \frac{\sum^k t_{ik}}{\sum^k (A_k/a_k) t_{ik}}$$

where t'_{ij} = future traffic flow from i to j

t_{ij} = existing traffic flow from i to j

P_i and A_j are total future trips produced by zone i and attracted to zone j respectively.

p_i and a_j are total existing trips produced by zone i and attracted to zone j respectively.

k = total zones.

The procedure must be iterated by substituting t'_{ij} for t_{ij}, $\sum_j t'_{ij}$ for p_i, $\sum_i t'_{ij}$ for a_j.

Agreement to between 1 and 5 per cent is achieved by successive iterations.

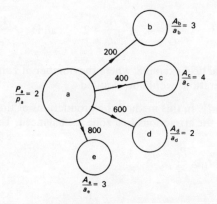

Figure 7.1 The Fratar method of trip distribution

The Fratar method can be illustrated by the simple example shown in figure 7.1 which shows the growth factors and the existing trip pattern.

Then

$$t'_{ij} = t_{ij} \times \frac{P_i}{p_i} \times \frac{A_j}{a_j} \times \frac{\sum^k t_{ik}}{\sum^k \left(\frac{A_k}{a_k}\right) t_{ik}}$$

that is

$$t'_{ab} = 200 \times 2 \times 3 \times \frac{(200 + 400 + 600 + 800)}{200 \times 3 + 400 \times 4 + 600 \times 2 + 800 \times 3}$$

$$= 414$$

$$t'_{ac} = 400 \times 2 \times 4 \times \frac{(200 + 400 + 600 + 800)}{200 \times 3 + 400 \times 4 + 600 \times 2 + 800 \times 3}$$

$$= 1103$$

$$t'_{ad} = 600 \times 2 \times 2 \times \frac{(200 + 400 + 600 + 800)}{200 \times 3 + 400 \times 4 + 600 \times 2 + 800 \times 3}$$

$$= 828$$

$$t'_{ae} = 800 \times 2 \times 3 \times \frac{(200 + 400 + 600 + 800)}{200 \times 3 + 400 \times 4 + 600 \times 2 + 800 \times 3}$$

$$= 1655$$

In this example the future trips produced by zone a meet the requirements that $P_a/p_a = 2$, but the requirements that

$$\frac{A_b}{a_b} = 3; \qquad \frac{A_c}{a_c} = 4; \qquad \frac{A_d}{a_d} = 2 \qquad \text{and} \qquad \frac{A_c}{a_c} = 3$$

are not met and in a practical example, where the number of zones would be considerably greater, further iterations would be required.

(d) *The Furness Method*[2]. In this method the productions of flows from a zone are first balanced and then the attractions to a zone are balanced. This may be expressed

$$t'_{ij} = t_{ij} \times \frac{P_i}{p_i}$$

$$t''_{ij} = t'_{ij} \times \frac{A_j}{\Sigma \text{ trips attracted to j in first iteration}}$$

$$t'''_{ij} = t''_{ij} \times \frac{P_{ij}}{\Sigma \text{ trips produced by i in second iteration}}$$

where the symbols are as previously stated.

Usually these simplified approaches to trip distribution are only suitable for smaller surveys where similar growth factors are applied to zones and where considerable areas of new development are not expected. The following extract from *Land Use/Transport Studies for Smaller Towns*[3] illustrates the use of growth factors and the Furness method of trip.

The four steps in the calculation are:

1. Total the outgoing trips for each zone and multiply by the zonal growth factor to obtain the predicted origin outgoing totals.
2. Multiply lines in the matrix by the appropriate origin factor.
3. Total the incoming trips into each zone and divide into the predicted incoming totals to obtain the destination factors.
4. Repeat the iteration processes until the origin or destination factor being calculated is sufficiently close to unity (within 5 per cent is normally satisfactory).

Example

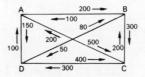

Figure 7.2 The Furness method of trip distribution (initial flows)

TABLE 7.1   Present flows

		Destinations				Present outgoing totals	Predicted outgoing totals	Growth (origin) factor
	Origin	A	B	C	D			
	1	2	3	4	5	6	7	8
9	A	0	200	500	150	850	2550	3·0
10	B	100	0	300	50	450	1125	2·5
11	C	200	200	0	300	700	1400	2·0
12	D	100	80	400	0	580	925	1·6
13	Present incoming totals	400	480	1200	500		6000	
14	Predicted incoming totals	480	720	3600	1200		6000	
15	Acceptance (destination) factor	1·2	1·5	3·0	2·4			

Notes: col. 7 = col. 6 x col. 8
line 14 = line 13 x line 15
Total col. 7 must equal approximately total line 14; that is future trip ends in system must balance, any adjustment to calculated acceptance factors necessary to secure balance being made at the non-residential end of the trips.

TABLE 7.2   Iteration 1

		A	B	C	D
		16	17	18	19
20	A	0	600	1500	450
21	B	250	0	750	125
22	C	400	400	0	600
23	D	160	128	640	0
24		810	1128	2890	1·175
25		480	720	3600	1200
26	New destination factors	0·59	0·64	1·25	1·02

Figures in this matrix are those in lines 9–12 multiplied by respective origin factors in column 8

As in line 14

TABLE 7.3   Iteration 2

	A	B	C	D		(as Col. 7)	
A	0	384	1870	458	2712	2550	0·94
B	148	0	935	128	1211	1125	0·93
C	217	256	0	612	1105	1400	1·26
D	95	82	797	0	974	925	0·95

Figures in this matrix are columns 16–19 multiplied by respective new destination factors on line 26.

TABLE 7.4   Iteration 3

	A	B	C	D
A	0	362	1760	432
B	138	0	870	119
C	300	324	0	775
D	90	78	752	0
	528	764	3382	1326
	480	720	3600	1200
	0·91	0·94	1·07	0·91

TABLE 7.5   Iteration 4

	A	B	C	D			
A	0	441	1882	392	2615	2550	0·98
B	125	0	906	108	1141	1125	0·99
C	273	305	0	702	1280	1400	1·09
D	82	73	807	0	962	925	0·96

TABLE 7.6   Iteration 5

	A	B	C	D
A	0	333	1840	383
B	123	0	895	107
C	299	334	0	770
D	79	70	777	0
	501	737	3512	1260
	480	720	3600	1200
	0·96	0·96	1·02	0·95

When all the origin or destination factors being calculated in one iteration are within 5 per cent of unity, the result may be considered satisfactory.

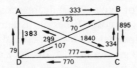

Figure 7.3 The Furness method of trip distribution (distributed flows)

General comment on growth factor methods

The usual application of growth factor methods is in updating recent origin–destination data where the time scale is short and where substantial changes in land use or communications are not expected or have not already taken place. The use of a growth factor method is largely dependent on the accurate calculation of the growth factor itself and this is a likely source of inaccuracy. It is however the lack of any measure of travel impedance which is the major disadvantage of these methods; without this it is impossible to take into account the effect of new and improved travel facilities or the restraining effects of congestion.

Trip distribution using synthetic models

The use of synthetic distribution models allows the effect of differing planning strategies and communication systems and, in particular, travel cost to be estimated, whereas growth factor methods base future predictions on the existing pattern of movement.

The models are usually referred to as synthetic models because existing data is analysed in order to obtain a relationship between tripmaking and the generation and attraction of trips and travel impedance. These models seek to determine the causes of present day travel patterns and then assume that these underlying causes will remain the same in the future. The most widely used trip distribution model is the so-called 'gravity model'. It has been given this name because of its similarity with the gravitational concept advanced by Newton. It states that trip interchange between zones is directly proportional to the attractiveness of the zones to trips and inversely proportional to some function of the spatial separation of the zones.

The gravity model may generally be stated as

$$t_{ij} = kA_iA_jf(Z_{ij})$$

where A_i and A_j are measures of the attractiveness of the zones of origin and destination to tripmaking and $f(Z_{ij})$ is some function of deterrence to travel expressed in terms of the cost of travel, travel time or travel distance between zones i and j.

If $f(Z_{ij})$ is taken as $1/Z_{ij}^2$ then the formula is similar to the law of gravitational attraction and for this reason the model is referred to as the gravity model.

Tanner[4] has shown that this form of the deterrence function cannot give valid estimates of travel over large or small distances. He suggested a function of the form

$$\exp(-\lambda Z_{ij}) \times Z_{ij}^{-n}$$

where λ and n are constants.

For work journeys a typical value of λ was found to be 0·2 where Z_{ij} was measured in miles. For many purposes n may be taken as 1.

As Z_{ij} increases the effect of $\exp(-\lambda Z_{ij})$ decreases and the effect of Z_{ij}^n increases.

This is the distribution of trip costs, times or distances which is observed in most studies of tripmaking. The trips that are least in cost, distance or time occur most frequently while trips that are greater in cost, distance or time are observed less frequently. There is considerable evidence that the value of the deterrence function varies with trip type and for this reason practically all studies use differing values for differing trip categories. It is also probable that the function varies with time, but this effect is at present largely ignored.

In practice tripmaking between zones is assumed to be proportional to the total trips generated by the zone of origin of the trip and the total trips attracted by the zone of destination of the trip, or

$$t_{ij} = KP_iA_jf(Z_{ij})$$

where K is a constant.

P_i is the total trip production of zone i, that is $\sum_j t_{ij}$

A_j is the total trip attraction of zone j, that is $\sum_i t_{ij}$

The value of K is frequently taken as

$$\frac{1}{\sum_j A_jf(Z_{ij})}$$

so that

$$t_{ij} = \frac{P_iA_jf(Z_{ij})}{\sum_j A_jf(Z_{ij})}$$

This is often referred to as a production constrained model because the use of this form of K results in the total trip production of zone i; that is $\sum_j t_{ij}$ is equal to P_i.

When it is desired to doubly constrain the distribution model, that is

$$\sum_j t_{ij} = P_i$$

$$\sum_i t_{ij} = A_j$$

then K may be replaced by $c_i d_j$, giving

$$t_{ij} = c_i d_j P_i A_j f(Z_{ij})$$

and

$$\sum_j t_{ij} = \sum_j c_i d_j P_i A_j f(Z_{ij}) \quad \text{or} \quad c_i = \frac{1}{\sum_j d_j A_j f(Z_{ij})}$$

and similarly

$$d_j = \frac{1}{\sum_i c_i P_i f(Z_{ij})}$$

The distribution is constrained by setting d_j equal to unity, and obtaining

$$t_{ij} = \frac{d_j P_i A_j f(Z_{ij})}{\sum_j d_j A_j f(Z_{ij})}$$

If $\sum_i t_{ij}$ is not equal to A_j then d_j is made equal to $A_j / \sum_i t_{ij}$ and the procedure iterated until $\sum_i t_{ij}$ approaches A_j.

Past experience has shown that the exponent of travel time varies with travel purpose and also with travel time. For this reason the gravity model is usually presented in the revised form

$$t_{ij} = \frac{P_i A_j F_{ij} K_{ij}}{\sum_{j=1}^{n} A_j F_{ij} K_{ij}}$$

where F_{ij} = an empirically derived travel time or friction factor which expresses the average area wide effect of spatial separation on trip interchange between zones which are z_{ij} apart. This factor approximates $1/Z^n$ where n varies according to the value of Z expressed as the travel time between zones.

K_{ij} = a specific zone-to-zone adjustment factor to allow for the effect on travel pattern of defined social or economic linkage not otherwise accounted for in the gravity model formulation.

Standard computer programs are available for determining the most suitable values of F_{ij} and K_{ij} for the particular transportation study. These values are then assumed to remain constant in the future to allow future flows to be predicted; this procedure is known as calibration.

The travel time factors are calculated on a survey wide basis by assuming a value of F_{ij} for a given time range. t_{ij} is calculated for all the flows within the same time range and compared with observed values. The procedure is iterated until agreement is . obtained and finally zone-to-zone agreement is obtained by the use of K factors.

Example of the use of a gravity model

Car driver work trips produced by the residents of zone 1 amount to 1000 trips. It is desired to distribute these trips to zones 1-4 which have the following characteristics.

Zone 1 has an intrazonal time of 7 minutes and has 1000 work trips attracted to it from all zones in the study area. (The intrazonal time is the average travel time of trips which have origins and destinations within the zone.) Terminal time 2 minutes.

Zone 2 is 15 minutes from zone 1 and has a total of 700 work trips attracted to it from all zones in the study area. Terminal time 3 minutes.

Zone 3 is 19 minutes from zone 1 and has a total of 6000 work trips attracted to it from all zones in the study area. Terminal time 2 minutes.

Zone 4 is 20 minutes from zone 1 and has a total of 3000 work trips attracted to it from all zones in the study area. Terminal time 3 minutes.

The travel time factors applicable to journeys of this type are given in table 7.7. The terminal time is the average time to park and walk to the destination, or walk from the origin to the car park.

TABLE 7.7

Travel time (minutes)	F
1	200
5	120
7	100
11	80
14	68
16	61
17	58
18	52
20	49
21	47
23	45
25	39

TABLE 7.8

1	2	3	4	5	6	
Zones	A_j	Terminal times	Travel time	t_{ij}	F_{ij}	column 2 x column 6
1-1	1000	2 + 2	7	11	80	80 000
1-2	700	2 + 3	15	20	49	34 300
1-3	6000	2 + 2	19	23	45	270 000
1-4	3000	2 + 3	20	25	39	117 000

Σ 501 300

Using the trip distribution equation

$$T_{ij} = \frac{P_i A_j F_{ij}}{\sum_j A_j F_{ij}}$$

where A_j is given in column 2
and F_{ij} is given in column 6

$$\text{Trips } 1\text{-}1 = \frac{1000 \times 80\,000}{501\,300} = 160$$

$$\text{Trips } 1\text{-}2 = \frac{1000 \times 34\,300}{501\,300} = 68$$

$$\text{Trips } 1\text{-}3 = \frac{1000 \times 270\,000}{501\,300} = 539$$

$$\text{Trips } 1\text{-}4 = \frac{1000 \times 117\,000}{501\,300} = 233$$

Application of the gravity model to the London Transportation Study

The gravity model concept was used for trip distribution in the London Transportation Study[12]. It was assumed that the number of trips generated in one district and attracted to another was a function of trip generation and attraction of the two districts respectively and the travel time between them.

These functions, referred to as distribution functions, were calculated from the survey data and assumed to remain constant with time. Trips were stratified by purpose and mode into six types: work, other home-based and non-home-based, for both car owner and public transport trips. Because these functions may vary with trip length the trips were divided into thirteen trip time-length intervals so that approximately equal numbers of trips were in each interval.

Using travel data from the household interviews the distribution functions F_{kt} for each district of attraction k and time interval t were calculated

$$F_{kt} = \frac{\sum\limits_{i(t)} T_{ik}}{A_k \sum\limits_{i(t)} G_i}$$

where i   = district of generation,

T_{ik} = trips generated in district i and attracted to district k,

A_k = total trips attracted to district k,

G_i = total trips generated in district i,

$\sum\limits_{i(t)}$ = summation over all districts of generation, falling in time interval t.

It was found that when these functions were plotted to a logarithmic scale against time, they showed an approximately linear relationship but with a discontinuity in gradient at a trip time length of around 25 minutes.

Forecasting of future trips was carried out separately for each trip purpose after the removal of a fixed percentage of intra-district trips. The first iteration of the distribution was obtained by the use of the model

$$T_{1ik} = \frac{F_{ik}A_kG_i}{\sum\limits_{k}(F_{ik}A_k)}$$

where F_{ik} = distribution function for trips from districts i to k using 1981 trip times,

A_k = forecast attractions in district k

G_i = forecast generations in district i

A ratio was then obtained for each district of attraction, so that

$$R_{1k} = \frac{A_k}{\sum\limits_{i} T_{1ik}}$$

Using this ratio, a second iteration was made

$$T_{2ik} = \frac{F_{ik}R_{1k}A_kG_i}{\sum\limits_{k}(F_{ik}A_kR_{1k})}$$

followed by the calculation of

$$R_{2k} = \frac{R_{1k}A_k}{\sum\limits_{i} T_{2ik}}$$

The procedure was carried out for a third time, the criterion for convergence being that the average (unsigned) error should be within the range 2 to 5 per cent.

Other synthetic methods of trip distribution

While the gravity model has found considerable application in the distribution of generated trips there have been attempts to use other mathematical models which reflect the motivations of tripmakers more closely. Other synthetic trip distribution models that have been used in transportation studies are:

(i)   the intervening opportunities model[5];
(ii) the competing opportunities model[6];

In the intervening opportunities model, which was first used in connection with the Chicago Area Transportation Study[7], the basic assumption is that all trips will remain as short as possible, subject to being able to find a suitable destination. The competing opportunities model has been evaluated in a study of travel patterns observed in the City of Lexington and Fayette County, Kentucky[8]. It differs from the intervening opportunities model in that the adjusted probability of a trip ending in a zone is the product of two independent probabilities, the probability of a trip being attracted to a zone and the probability of a trip finding a destination in that zone.

The intervening opportunities model may be derived as follows

$$t_{ij} = A_i [\exp(-LD) - \exp(-LD_j + 1)],$$

and the probability that a trip will terminate by the time D possible destinations have been considered is

$$P(D) = 1 - \exp(-LD)$$

Then t_{ij} is equal to P_i multiplied by the probability of a trip terminating in j.

In this treatment L is a constant representing the probability of a destination being accepted, if it is considered,

D_j is the sum of all possible destinations, considered in order of travel cost, for zones between i and j, but excluding j,

$D_j + 1$ is the sum of all possible destinations for zones, considered in order of travel cost, between i and j, including j.

The model is calibrated by rearranging the expression for $P(D)$ to give

$$1 - P(D) = \exp(-LD)$$

Taking logarithms to the base e of both sides

$$\ln(1 - P(D)) = -LD$$

From the trip pattern the value of L can be determined by regression and L adjusted until sector to sector accuracy is achieved.

Difficulties with this method are: firstly, that L values differ for long and short trips and for this reason trips have to be stratified by trip cost (length or time); secondly, as the number of destination opportunities increase in the future, then the number of shorter trips will increase. For this reason it is usually necessary to maintain the same proportion of shorter trips in the future as are observed in the present pattern; thirdly, it is theoretically necessary to have an infinite number of destinations to utilise all trip origins. Ruiter has indicated a model revision to overcome this difficulty[9].

A comparative study of trip distribution methods carried out by Heanue and Pyers[10] suggested that the intervening opportunities model gave slightly less accurate results than the gravity model in base year simulation for Washington. The gravity model did however use socio-economic adjustment factors and without these the opportunity model was better than the unadjusted gravity model.

Blunden[11] has given the form of the competing opportunities model as

$$t_{ij} = P_i \frac{A_j}{\sum\limits_j A_j} \bigg/ \sum \left(\frac{A_j}{\sum\limits_j A_j} \right)$$

where zonal trip attractions are summed in the order of their travel cost (time or distance) from the origin i.

Lawson and Dearinger[8] have used this model for the distribution of work trips in the City of Lexington and Fayette County. The model is difficult to calibrate and use and they concluded in a comparative study of trip distribution methods that the gravity model produced the best correlation with the existing trip pattern.

References

1. T. J. Fratar. Vehicular trip distributions by successive approximations. *Traff. Q.*, 8 (1954), 53–64
2. K. P. Furness. Time function iteration. *Traff. Engng Control*, 7 (1965), 458–60
3. Ministry of Transport. Land Use/Transport Studies for Smaller Towns. *Memorandu to Divisional Road Engineers and Principal Regional Planners*. London (1965)
4. J. C. Tanner. Factors affecting the amount of travel. *DSIR Road Research Technic Paper* No. 51. H.M.S.O. (1961)
5. S. A. Stouffer. Intervening opportunities–a theory relating mobility and distance. *Am. soc. Rev.*, 5 (1940), 347–56
6. A. R. Tomazinia and G. Wickstrom. A new method of trip distribution in an urban area. *Highw. Res. Bd Bull.* 374 (1962), 254–7
7. Chicago Area Transportation Study, Final Report, 11 (1960)
8. H. C. Lawson and J. A. Dearinger. A comparison of four work trip distribution models. *Proc. Am. Soc. civ. Engrs*, 93 (November 1967), 1–25
9. F. R. Ruiter. Discussion on R. W. Whitaker and K. E. West. The intervening opportunities model: a theoretical consideration. *Highw. Res. Rec.* 250, Highway Research Board (1968)
10. K. E. Heanue and C. E. Pyers. A comparative evaluation of trip distribution procedures. *Highw. Res. Rec.* 114, Highway Research Board (1966)
11. W. R. Blunden. *The Land-Use/Transport System*, Pergamon, Oxford (1971)
12. Greater London Council. Movement in London. County Hall, London (1969)

Problems

1. Select one or more correct answers to the following questions:

Trip distribution is an element of the process by which person movement is synthesisec in which:

(a) trips from each zone of the transportation study area are calculated and their path through the transportation network determined;

(b) the destinations of trips with known origins are determined;

(c) a known pattern of trips is increased in accordance with factors determined from anticipated conditions at both the zones of origin and destination of the trips.

2. It is anticipated that substantial growth will take place in several zones of a transportation study and that in addition some of the growth will take place in areas which do not at the present day generate any trips. Which of the following methods of trip distributions would be appropriate?

(a) a method based on the average growth of the zone of origin and of the zone of destination of the trip;

(b) an iterative method in which the future attraction and production of trips is correctly assessed by the alternate use of an attraction factor and a production factor;

(c) a method based on the assumption that trip interchange between two zones is proportional to the total trips generated by the zone of origin and the total trips attracted by the zone of destination and inversely proportional to a function of the cost of travel between the zones.

3. The present and the future generated and attracted trips from four traffic zones of a transportation study are as given below, together with the present trip matrix.

TABLE 7.9   Present and future generated and attracted trips

Zone	A	B	C	D
present generated trips	1500	900	1800	800
present attracted trips	1200	1000	1500	2000
future generated trips	3000	1200	2700	2400
future attracted trips	1800	3000	3500	4000

TABLE 7.10   Present trip matrix

Origins	Destinations			
	A	B	C	D
A	–	400	400	300
B	200	–	300	200
C	400	300	–	600
D	200	100	300	–

(a) Calculate the first approximation to the future trips between the zones using the average factor method.

(b) Calculate the second approximation to the future trips between the zones using the Furness method.

4. Trips between the traffic zones of a proposed new town are assumed to be proportional to the trips generated by the zone of origin and the trips attracted by the zone of destination of the trip and inversely proportional to the 2nd power of the travel time between the zones. Details of three traffic zones are given in table 7.11 and the value of the future trips from C to A is also given in table 7.12.

TABLE 7.11

Zone	Attracted trips	Generated trips
A	2400	3600
B	1600	2000
C	4000	5000

The travel time between the zones is 10 minutes.

TABLE 7.12

		Zone of destination		
		A	B	C
zone of	A			X
origin	B	Y		
	C	208	Z	

What is the correct value of X, Y and Z in table 7.12?

Solutions

1. Trip distribution is an element of the process by which person movement is synthesised in which (b) the destinations of trips with known origins are determined, or (c) a known pattern of trips is increased in accordance with factors determined from anticipated conditions at both the zones of origin and destination of the trips.

2. When substantial growth will take place in several of the zones of a transportation study area and in addition there will be growth in previously undeveloped areas then the method of trip distribution will be (c) a method based on the assumption that trip interchange between two zones is proportional to the total trips generated by the zone of origin and the total trips attracted by the zone of destination and inversely proportional to a function of the cost of travel between the zones.

3. (a) Calculation of the trips between the zones using the average factor method.

Zone	A	B	C	D
Attraction factor	1·5	3·0	2·3	2·0
Production factor	2·0	1·3	1·5	3·0

then

$$t'_{AB} = 400 \frac{(2 \cdot 0 + 3 \cdot 0)}{2} = 1000$$

$$t'_{AC} = 400 \frac{(2 \cdot 0 + 2 \cdot 3)}{2} = 860$$

$$t'_{AD} = 300 \frac{(2 \cdot 0 + 2 \cdot 0)}{2} = 600$$

$$t'_{BA} = 200 \frac{(1 \cdot 3 + 1 \cdot 5)}{2} = 280$$

$$t'_{BC} = 300 \frac{(1 \cdot 3 + 2 \cdot 3)}{2} = 540$$

$$t'_{BD} = 200 \frac{(1 \cdot 3 + 2 \cdot 0)}{2} = 330$$

$$t'_{CA} = 400 \frac{(1 \cdot 5 + 1 \cdot 5)}{2} = 600$$

$$t'_{CB} = 300 \frac{(1 \cdot 5 + 3 \cdot 0)}{2} = 675$$

$$t'_{CD} = 600 \frac{(1 \cdot 5 + 2 \cdot 0)}{2} = 1000$$

$$t'_{DA} = 200 \frac{(3 \cdot 0 + 1 \cdot 5)}{2} = 450$$

$$t'_{DB} = 100 \frac{(3 \cdot 0 + 3 \cdot 0)}{2} = 300$$

$$t'_{DC} = 300 \frac{(3 \cdot 0 + 2 \cdot 3)}{2} = 795$$

These future trip interchanges can now be tabulated in an origin–destination matrix.

Origins	Destinations			
	A	B	C	D
A		1000	860	600
B	280		540	330
C	600	675		1000
D	450	300	795	

If complete details of the trip interchanges between all the zones of the survey had been given it would be possible to note that the future attractions and generations of the zones as obtained in this first iteration did not agree with the values given in table 7.9. It would then be necessary to carry out the procedure again using new values of attraction and production factors based on the ratio of first iteration interchanges to future interchanges.

(b) Calculation of the trips between the zones using the Furness method.

Iteration 1
Using the production factors calculated in (a)

	A	B	C	D
A	–	800	800	600
B	260	–	390	260
C	600	450	–	900
D	600	300	900	–
Attractions	1460	1550	2090	1760
Required future attractions	1200	2400	2300	2200
New attraction factor	0·82	1·55	1·10	1·25

Iteration 2

	A	B	C	D
A	–	1240	880	750
B	213	–	429	325
C	492	698	–	1125
D	492	475	990	–

4. It is stated that

$$t'_{ij} = \frac{KP_iA_j}{t^2}$$

From the data given of the trip interchange from C to A

$$208 = \frac{K \times 5000 \times 2400}{10^2}$$

$$K = \frac{208}{120\,000}$$

Then

$$X = t'_{AC} = \frac{208 \times 3600 \times 4000}{120\,000 \times 100}$$

$$= 250 \text{ trips}$$

$$Y = t'_{BA} = \frac{208 \times 2000 \times 2400}{120\,000 \times 100}$$

$$= 83 \text{ trips}$$

$$Z = t'_{CB} = \frac{208 \times 5000 \times 1600}{120\,000 \times 100}$$

$$= 139 \text{ trips}$$

8

Modal split

Trips may be made by differing methods or modes of travel and the determination of the choice of travel mode is known as modal split. In the simplest case when a small town is being considered the choice is normally between one form of public transport and the private car, with the car being used for all trips where it is available. In such a situation most trips on the public transport network are captive to public transport and very little choice is being exercised. In the larger conurbations however the effect of modal split is of very considerable significance and is greatly influenced by transport policy decisions.

Modal split should not be viewed as an entity; it is closely related in the real situation with trip generation and distribution. It has been shown in chapter 6 that additional tripmaking occurs when a private car is available and it has been observed that the destination is often influenced by the ease with which the car can be used. If the use of the private car is restricted then it is likely that the number of trips generated will be decreased rather than be made by an alternative travel mode.

In the simulation of the real system which is referred to as the transportation planning process, modal split may be carried out at the following positions in the process.

(a) Modal split may be carried out as part of trip generation whereby the number of trips made by a given mode are related to characteristics of the zone of origin. This means that transport trips are generated separately from private transport trips.

(b) Modal split may be carried out between trip generation and distribution. Car owning households in the zone of origin have a choice of travel mode depending upon the car/household ratio while non-car owning household trips are captive to public transport.

(c) Modal split may be carried out as part of the trip distribution process relating distribution not only to travel time by mode but also with functions including the relative elasticity of demand with respect to tripmaking characteristics by the available modes.

(d) Modal split may be carried out between the trip distribution and the trip assignment process. Trip distribution allows journey times both by public and private transport to be estimated and then the modal split between public transport tr ps may be made on the basis of travel time and cost.

These varying approaches will now be considered in greater detail.

53

(a) Modal split considered as part of the trip generation process

The direct generation of public and of private transport trips was used in early work in the USA and has often been carried out in surveys of small urban areas in the UK. The Leicester transportation plan is an example of this approach where regression equations are developed for several trip types by the four modes, car driver, bus passenger, car passenger and other modes. Usually the modal split is made on the basis of car ownership in the zone of origin, distance of the zone of origin from the city centre and residential density in the zone of origin. Sometimes the relative accessibility of the zone of origin to public transport facilities is also included.

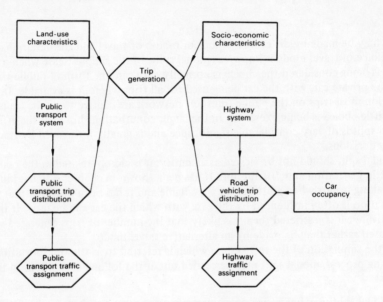

Figure 8.1 A generalised trip end modal-split procedure

This approach makes it difficult to take into account changes in the public transport network, improvements in the highway system and the restraint of private car use by economic means. Usually these models indicate a very high future car use and arbitrary modal split has to be imposed after the assignment process.

Figure 8.1 illustrates the transport planning procedure graphically when public transport and car trips are generated separately.

(b) Modal split carried out between trip generation and distribution

In this approach person trips are predicted and the percentage of these trips made by public and private transport estimated from such factors as socio-economic and land-use characteristics, the quality of the public transport system and the number of cars available. The assumption is made in this method that the total number of trips generated is independent of the mode of travel.

A typical example of this technique is given in *Transportation and Parking for Tomorrow's Cities*[1] where the modal split decision is made entirely on the basis of the average persons per car in the zone of origin. A generalised mapping of the function is shown in figure 8.2. It is to be expected that socio-economic, land use and the availability of public transport facilities would modify the form of the function and similar mappings have been used showing the relationship between percentage of trips by public transport and income.

Regression analysis allows the various factors that influence modal split to be incorporated into the analysis of existing behaviour but it is necessary not only for the regression analysis to explain the existing modal choice of tripmakers, but also to be sensitive to the factors that will affect modal choice in future projections.

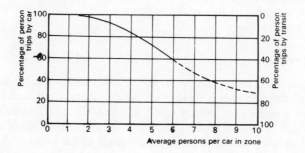

Figure 8.2 Relationship between car use and car ownership (adapted from ref. 1)

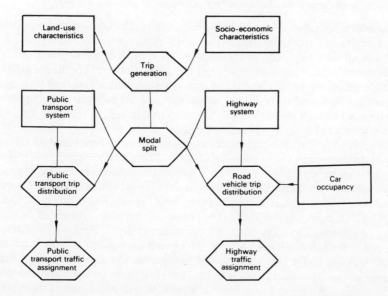

Figure 8.3 The generalised model when modal split is carried out between generation and assignment

In the Baltimore transportation study[1] Wilbur Smith and Associates developed the following regression equation to determine modal choice

$$T = 1 \cdot 29 X_1 - 1 \cdot 036 X_2 + 0 \cdot 529 X_3 + 0 \cdot 150 X_4 - 102$$

where T = number of public transport trips $(R = 0 \cdot 93)$

X_1 = labour force in generating zone

X_2 = no. of cars in generating zone

X_3 = public transport service index

X_4 = no. of students in zone

The transport planning procedure when modal split is carried out between generation and assignment is illustrated graphically in figure 8.3.

(c) Modal split carried out as part of the trip distribution process

An attempt to simulate human behaviour more closely and overcome difficulties inherent in the other methods is made when trip distribution and modal split are carried out at the same time for each journey made. The SELNEC study embodies this approach in which two travel modes compete at the destination end of the trip. Further details of this approach can be obtained from a study of the works of A. G. Wilson, A. F. Hawkins, D. J. Wagon and others[2,3].

(d) Modal split carried out between trip distribution and assignment

This approach is frequently used in transportation studies because it allows the cost and level of service of a trip to be used as the modal split criterion.

Because of the complexity of the transportation process, travel time alone is often used as the cost criterion. Normally travel times based on road speeds are utilised to distribute the choice trips. These travel times together with travel time by public transport are then used to determine modal split and the public transport portion of these trips is added to the captive public transport trip ends as shown in figure 8.4, which is abstracted from Movement in London[4].

The simplest modal split criteria use diversion curves that are based on travel time ratios or differences and have been obtained for a considerable number of cities during transportation surveys. Figures 8.5 and 8.6 reproduced from Transportation and Parking for Tomorrow's Cities illustrate the form of the curve obtained. Individual curves vary because many other factors besides the time ratio are involved but they all show decreasing public transport use as the travel time ratio increases. An added refinement is the use of differing diversion curves for differing trip types. Figure 8.7 illustrates this approach in the London Travel Survey. While such a simplified approach can take into account highway congestion via increased travel times and also the quality of the public transport service by reduced transit times, it does not take into account restraint factors such as parking charges and road pricing. More effective simulation of the real system is obtained if the time ratio is replaced by a cost ratio and there is evidence that cost differences represent person choice more realistically.

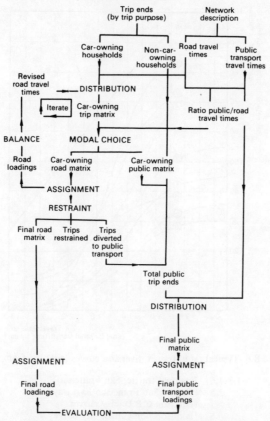

Figure 8.4 Procedure when modal split is carried out between trip distribution and assignment

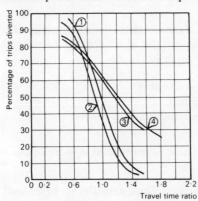

Figure 8.5 Typical public transport and highway diversion curves (adapted from ref. 1)

1   U.S. Bureau of Public Roads
2   American Association of State Highway Officials
3   San Francisco Transportation Committee
4   Chicago Transportation Usage Study

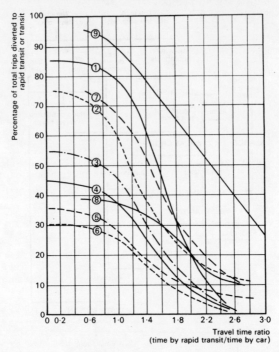

Figure 8.6 Typical rapid transit diversion curves (adapted from ref. 1)

1,3,4,6   Regional traffic, San Francisco
2,5   Internal San Francisco East Bay traffic
7   Hamilton C.B.D. city curves
8   Hamilton non-C.B.D. city curves
9   Toronto work trips

More complex modal choice models are usually obtained by regression analysis and discriminate analysis. The variables which are used to estimate modal choice may be classified into the following three types:

(a) characteristics of the tripmaker;
(b) characteristics of the trip;
(c) characteristics of the transportation system.

These characteristics will now be considered in more detail.

(a) *Tripmaker characteristics*. Three variables are often used:

car ownership;
income;
net residential density.

(N.B. These are highly correlated independent variables.)

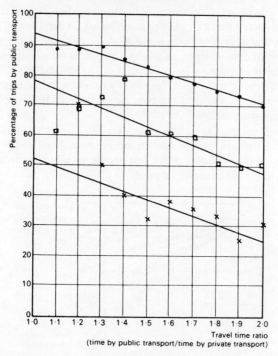

Figure 8.7 The London Traffic Survey, specimen diversion curves (adapted from ref. 4)

work trips O
other home-based trips □
non-home-based trips X

(b) *Trip characteristics.* It is usual to develop differing modal split relationships for differing trip types, namely:

(i)   home–work;
(ii) home–school;
(iii) home–social purposes etc.

for peak and non-peak periods.

(c) *System characteristics.* It is usual to assess the cost of travel by competing mode in the simplest way by time cost, as in diversion curve use. Difficulties arise when it is required to value time and many studies have been carried out to find how much people are prepared to pay to save travelling time. Current work indicates that this may be in the region of 30 per cent of the wage rate for the individual.

Variations in modal split characteristics with these factors are illustrated by the London Travel Survey in which 54 diversion curves were obtained from the survey data. These comprise 3 trip purposes (work, other home-based, non-home-based), long and short trips (divided at 15 minute travel time by private transport excluding parking) for each of the nine differing sectors of the transportation study area. Specimen diversion curves are shown in figure 8.7.

60 TRAFFIC ANALYSIS AND PREDICTION

A very detailed approach to modal split has been made by the Traffic Research Corporation[5]. This was a multivariate approach based on income and service level considerations. A diversion curve procedure related modal split to five selected variables: the ratio of door-to-door travel time by public transit to that by automobile; the ratio of excess travel time by transit to that of the automobile (the service ratio); the ratio of out of pocket travel costs (excluding time) by public transit to that by automobile; the economic status of the tripmaker (median family income); and trip purpose (work, non-work, school).

Curves for five economic status levels are given in figure 8.8 and as time ratios become less favourable to transit, the modal split becomes inversely related to income level. These curves illustrate the obvious point that people with higher incomes are less apt to use transit, particularly when service is poor.

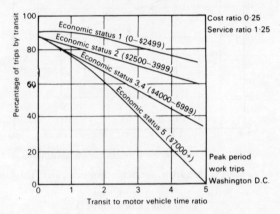

Figure 8.8 Modal split variations as developed by the Traffic Research Corporation (adapted from ref. 5)

The calibration of a model of this type, which has 40 diversion curves, requires a considerable amount of census information and additional details that are not always available from a standard transportation survey.

Another approach to the question of modal split has been made by the use of discriminate analysis by D. A. Quarmby[6].

An attempt is made to separate groups of tripmakers into different mode users using probability methods based on a combination of variables relating both to the characteristics of the journey and to the household from which the tripmakers come.

Just as category analysis has found use in trip generation, disaggregate modal split models have been developed to predict modal split. In a study carried out by Wilbur Smith and Associates in San Juan, total person trips were stratified by automobile ownership class and trip purpose.

There were three car ownership classes, 0, 1 and 2+, and two trip types, work and non-work. Trips were generated by purpose and automobile ownership class, distributed using separate gravity models, split into automobile and transit modes, and assigned to their respective systems.

However complex the modal choice mechanism employed it should not be forgotten that the ultimate goal is the simulation of human behaviour and the following quotation from *Transportation and Parking for Tomorrow's Cities* should be considered by all engaged in the simulation of future transport systems.

Finally, modal split must be kept in proper perspective. The area in which reasonable choices can be offered—where one form of transportation can effectively complement the other—is extremely limited. In most cities, transit's primary importance is for downtown travel. However, transit riders and central business district trip makers become smaller parts of the total travel market as living standards rise, urban areas expand and locational opportunities increase. Despite transit's importance in serving downtown trips, the greatest proportion of urban travel growth will be focussed elsewhere, directed largely to private vehicular transportation.

References

1. Wilbur Smith and Associates. *Transportation and Parking for Tomorrow's Cities*. New Haven, Connecticut (1966)
2. A. G. Wilson. The use of entropy maximising models. *J. transp. Econ. Policy*, 3 (1967), 108–25
3. A. G. Wilson, D. J. Wagon, E. H. E. Singer, J. S. Plant and A. F. Hawkins. The SELNEC transport model. Urban Studies Conference (1968)
4. Greater London Council. Movement in London. County Hall, London (1969).
5. T. Deen, W. Mertz and N. Irwin. Application of a modal split model to travel estimates for the Washington area. *Highw. Res. Bd Rec.* 38, Highway Research Board (1963)
6. D. A. Quarmby. Choice of travel mode for journey to work: some findings. *J. transp. Econ. Policy*, 1 (1967), 273–314

Problems

1. An equation of the following form predicts the number of trips on the public transport system

$$T = A + Bx_1 + Cx_2 + Dx_3$$

where T is the number of public transport trips,
   A, B, C and D are constants,
   x_1 is the number of generated trips in the zone of origin,
   x_2 is a measure of the public transport accessibility,
   x_3 is a measure of the car ownership in the zone of origin.
State whether this equation is used for the prediction of modal split after:

(a) the generation of trips;
(b) the distribution of trips between the zones;
(c) the assignment of trips to the network.

2. Trips have been distributed between the zones but they have not as yet been assigned to the road or the public transport network.

(a) Show graphically the usual relationship between the ratio of public to private transport travel time and the percentage use of public transport.
(b) State the factors which could be expected to modify the graphs.

3. Are the following statements true or false?

(a) Modal split is the division of trips between competing travel modes.

(b) When the modal split procedure is carried out as part of the trip generation procedure then the division between public and private trips is made on the basis of relative travel times.

(c) The trips generated in a zone may be divided between public and private trips on the basis of car ownership and bus accessibility in the zone of origin.

(d) Modal split may be carried out after the distribution of trips using the type of trip as a criterion.

Solutions

1. An equation of this form is used to carry out the modal split procedure after the generation of trips. Subsequently the trips are distributed between the zones and then assigned to either the public transport or the highway network.

2. (a) When the zone-to-zone movements are known the modal split procedure may be carried out on the basis of the relative cost of movement by private car or public transport. The relationship between the ratio of transport costs, expressed in terms of travel time, for private and public transport and the percentage use of public transport is illustrated in figure 8.9.

(b) It would be expected that a graph of this form would be modified by a considerable number of factors which include car ownership, public transport accessibility, relative income levels, trip length and purpose.

3. (a) Correct.

(b) Incorrect. When modal split is carried out as part of the trip generation procedure the division between competing travel modes is determined by the characteristics of the

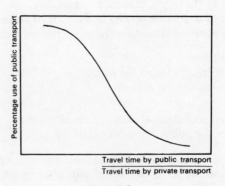

Figure 8.9

zone of origin of the trips. It cannot be made on the basis of relative travel times because the destination of trips has not yet been determined.

(c) Correct.

(d) Incorrect. When modal split is carried out after trip distribution then it is usually made on the basis of the relative travel costs by the differing modes of travel.

9

Traffic assignment

Previously the estimation of generated trip ends has been discussed together with the distribution of trips between the traffic zones. Modal split methods also have been reviewed in which the proportion of trips by the varying travel modes are determined. At this stage the number of trips and their origins and destinations are known but the actual route through the transportation system is unknown. This process of determining the links of the transportation system on which trips will be loaded is known as traffic assignment.

Apart from the very largest transportation surveys traffic assignment tends to deal with highway traffic. This is because it is usually not difficult to estimate the route taken by public transport users and also because the loading of trips on the public transport network does not materially affect the journey time.

The usual place of assignment in transportation planning synthesis is illustrated in the flow chart for Phase III of the London Transportation Study[1] (figure 8.4). Trip ends where there is no choice of travel mode, that is from non-car-owning households, are accumulated as public transport trip ends. Choice trips where a car is available are separated by the modal choice procedure into car trips and public transport trips, the public transport trips being accumulated and the car trips assigned to the network.

Usually it will then be found that the proposed road network is overloaded and some car trips will need to be restrained. If a car cannot be used then some trips will not be made at all, while other trips will be transferred to public transport and accumulated.

As the basis of assignment is usually travel time the travel times on the network links will vary the imposed loading. In addition as travel time is used in the trip distribution process it is necessary to carry out an iterative procedure between distribution, modal choice and assignment as illustrated in figure 8.4.

The change in speed with volume on a highway link is carried out using speed flow relationships for the varying highway types and it is interesting to consider just what are the effects of a speed change. Firstly it affects the choice of route because assignment is made on the basis of travel times through the network. Secondly it affects the destinations of trips because trips are distributed to varying destinations on the basis of travel time when a gravity model is used. Finally it may affect the choice of travel mode because modal choice is often made by a comparison of travel times.

There are many problems associated with speed/flow relationships, considerable variation being observed between differing highways even of the same type. There is also the additional problem that most transportation studies are based on 24 hour flows so that it is necessary to know the hourly variation and the directional distribution of flows.

In assignment it is first necessary to describe the transport network to which trips are being assigned. The network is described as a series of nodes and connecting links; in a highway network the nodes would be the junctions and the links the connecting highways. Centroids of traffic zones, at which it is assumed that all zonal trips are generated and to which they are attracted, are either at nodes or connected to them by additional links. The cost of using a link and a junction, usually in the form of travel times and delays, is given on the basis of the review of transport facilities carried out during the initial stages of the transportation survey.

There are four methods by which the assignment may be made. These are:

1. All-or-nothing assignment.
2. Assignment by the use of diversion curves.
3. Capacity restrained assignment.
4. Multipath proportional assignment.

All-or-nothing assignment

In this method an algorithm is used to compute the route of least cost, usually based on travel time between all the zone centroids. For each zone centroid selected as origin, a set of shortest routes from the origin to all the other zone centroids is referred to as a minimum tree. When the trips between two zones are assigned to the minimum path between the zones, then the assignment is said to take place on an all-or-nothing basis.

There are obvious difficulties with such a simplified approach, some of which are inherent in the other assignment methods. It is obviously incorrect to assume that all trips commence and terminate at a zone centroid. If the length of the links within the zones is small compared with the length of remainder of the minimum link path, then the errors may not be so serious. Because of its simplicity, travel time is usually employed as a measure of link impedance, but travel times may not be precisely estimated by the traveller. The use of a cost function which reflects the perceived cost of travel is desirable. The loading on a link in this method is extremely sensitive to estimated link and node costs. If these have been incorrectly estimated, then the resulting assignment is open to question. There is also the problem that links with small travel costs will attract trips without any adjustment in link cost.

Assignment by the use of diversion curves

Originally diversion curves were used to estimate the traffic that would be attracted by a single new route or transport facility. It is thus necessary to compare travel cost with and without the new transport facility, the decision as to whether a trip would use the new facility being based on a cost ratio or difference between the with and the without situation. When diversion curves are used for assignment it is necessary to consider the network with and without the new facility or highway link as input to an all-or-nothing assignment. The travel costs from the two networks are then used with the diversion curve to determine the proportion of the trips that are diverted from the existing network and transferred to the network with the new facility.

An early application of this technique was used in the traffic studies for the first section of the M1 motorway[2] where 50 per cent of trips were diverted to the motorway if the cost (time) of the trip on the existing route was within ±10 minutes of the trip cost on the motorway. It has often been considered that differences of trip cost more effectively modelled human behaviour than ratios. A diversion curve of this type is illustrated in figure 9.1.

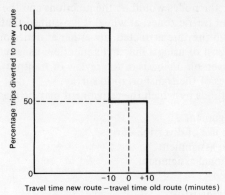

Figure 9.1 Early diversion curve

It is now usual to employ diversion curves similar to one shown in figure 9.2, which has been derived from observation of driver behaviour[3].

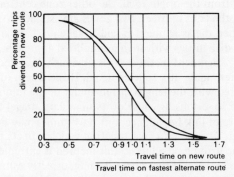

Figure 9.2 Specimen highway diversion curves (adapted from ref. 3)

In the USA indifference curves have been developed that take into account both time and distance savings. Figure 9.3 shows the form of the relationship proposed by Moskowitz[4] and based on observations of traffic diversion to new highways.

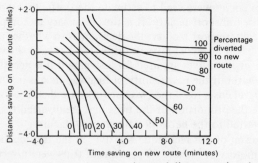

Figure 9.3 Highway diversion curves based on time and distance savings (adapted from ref. 4)

The mathematical form of the relationship is

$$\text{Percentage diverted to new route} = 50 + \frac{50(d + 0 \cdot 5t)}{\sqrt{[(d - 0 \cdot 5t)^2 + 4 \cdot 5]}} \qquad (9.1)$$

where d = distance saving on new route
t = time saving on new route.

Drivers perceive costs but as they are able to measure speed and form an assessment of journey speed, speed can be used to assess the diversion of drivers to a new route. This may well be true of the present motorway system in this country in that drivers are reasonably certain of being able to maintain a given speed but for certain journeys

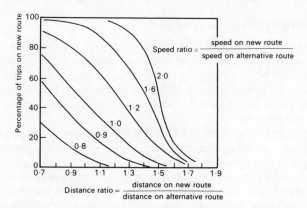

$$\text{Speed ratio} = \frac{\text{speed on new route}}{\text{speed on alternative route}}$$

$$\text{Distance ratio} = \frac{\text{distance on new route}}{\text{distance on alternative route}}$$

Figure 9.4 Highway diversion curves based on distance and speed ratios (adapted from ref. 5)

may not be aware if their journey time is greater on the new route. The diversion relationship shown in figure 9.4 developed for the Detroit Area Traffic Survey[5] illustrates the form of the assignment curves.

Capacity restrained assignment

All-or-nothing assignment results in a link with a favourable cost attracting a considerable number of trips, while links with unfavourable costs attract few trips. In practice this would result in the originally favourable link becoming overloaded, a situation which would not occur in real life.

The problem usually occurs when assigning trips to a highway network because in practice a balance exists between travel costs and flows. A number of routes between an origin and a destination are selected by tripmakers so that the perceived travel cost is approximately equal on all the routes. With a public transport network the choice of routes is more limited and the major difficulties occur when incorporating walking, waiting and interchange times.

Capacity-restrained assignment to highway networks attempts to tackle this problem by taking into account the relationship which exists between speed and flow on a highway. These relationships are discussed fully when considering the evaluation of transportation proposals.

There are several basic approaches to this type of assignment which have been used for capacity restrained assignment programs. The Bureau of Public Roads Model[6] first makes a complete all-or-nothing assignment to the network. The journey cost, usually expressed as time, is then updated on the basis of the flow assigned to it, and the whole procedure repeated for several iterations until the link costs show only a limited change on each iteration.

An alternative non-iterative method used in the Chicago Area Transportation Study[7] required the selection of a random origin zone, the calculation of the minimum tree and the assignment of flows from this zone to the tree. Journey costs on the links were then revised using the actual link flows and the procedure repeated with another randomly selected zone of origin.

Multipath proportional assignment

In urban areas there are many alternative routes between a given origin and destination and in actual fact tripmakers would be distributed over all these routes. This is because tripmakers will be unable to judge the route of least cost accurately and different trip-makers will make differing decisions. Multipath proportional assignment attempts to simulate this situation by assigning proportions of the trips between any two zones to a number of alternative routes.

Burrell[8] has described a model in which it is assumed that the tripmaker does not know the actual cost of using a link but that he associates a supposed cost with each link. This supposed cost is randomly generated from a distribution with the actual link cost as its mean and a given standard deviation representing the inability of a tripmaker to judge costs effectively. The assignment for one zone is then made on an all-or-nothing basis using the randomly determined link costs. Assignment of trips from the next zone considered will then be made on the basis of a further set of randomly determined link costs and the process repeated until all trips have been assigned to the network.

It has been reported that the assignment made by this model produces a satisfactory comparison with actual network flows giving more satisfactory results than all-or-nothing assignment.

Public transport assignment

Public transport assignment differs from the assignment of highway trips in that the public transport system is composed of a system of fixed services following stated routes at regular intervals. Because of the greater complexity of the assignment process the trips are usually assigned on an all-or-nothing basis. In the case of the public transport network the journey times that form the basis of the assignment must include walking, waiting and link travel times in addition to waiting time caused by the necessity to change from one route to another.

An assignment program developed for public transport and used for the London Transportation Study is Transitnet[9] developed by Freeman, Fox, Wilbur Smith and Associates. The program describes the network by routes. Each route has a frequency, nodes at which a transfer to either a walk link or a centroid connector can take place and a travel time between nodes. Waiting time values that are functions of the frequencie of the routes are used and there are time penalties built in to model a traveller's reluctan to change route or mode.

References

1. Greater London Council. Movement in London. County Hall, London (1969)
2. Road Research Laboratory. The London–Birmingham Motorway–traffic and economics. *Tech. Pap. Rd Res. Bd*, 46, H.M.S.O. (1961)
3. R. E. Schmidt and M. E. Campbell. Highway traffic estimation. Eno Foundation for Highway Traffic Control, Saugatuck, Connecticut (1956)
4. K. Moskowitz. California method of assigning diverted traffic to proposed freeways. *Highw. Res. Bd Bull.* 130 (1956)
5. Detroit Metropolitan Area Traffic Study, Detroit, Michigan (1956)
6. Bureau of Public Roads, Traffic Assignment Manual, Washington (1964).
7. J. D. Carroll. A method of traffic assignment to an urban network. *Highw. Res. Bd Bull.* 224 (1959)
8. J. Burrell. Multiple route assignment and its application to capacity restraint. 4th International Symposium on the Theory of Traffic Flow. Karlsruhe (1968)
9. R. B. Hewing and M. L. H. Hoffman. An explanation of the Transitnet algorithm. *Research Memorandum* 198. G.L.C. Department of Highways and Transportation (1969)

Problems

Are the following statements true or false?

1. Trip distribution determines trip destinations but traffic assignment determines the actual route taken to reach that destination.

2. Choice trips are those in which the tripmaker is under no compulsion to make the trip and will only do so if the trip length is small.

3. In the future it can be expected that tripmaking by the mode 'drive own car' will be restrained. This will result in all these trips being made by the public transport system.

4. One of the difficulties in the synthesis of the tripmaking process is that traffic assignment is frequently based on relative travel times. As the highway transport networks become more heavily loaded travel times are decreased and it is then necessary to consider the effect of this change on the trip distribution procedure.

5. Where a single new route is to be constructed then the process of determining which vehicles will choose to use the new route is known as traffic diversion. When however the effect of the construction of the new route on the whole highway network is being investigated then the traffic assignment process is used.

6. A problem of the all-or-nothing assignment technique is that a particular route will attract trips because it offers a cost saving that the average tripmakers would find difficult to perceive.

7. After the traffic assignment process has been completed the routes in the highway network carrying the larger traffic flows are output. This is the minimum highway network required and is referred to as the 'minimum tree'.

8. It is generally accepted that in most future large urban areas it will not be possible to construct a highway network capable of accommodating all the desired car trips. The proposed solution is to reduce the number of car trips allowed and their assignment to the highway network is known as capacity assignment.

9. Where there are many acceptable alternative paths between an origin and a destination then it is unreasonable to expect all the trips to go by the shortest route. The type of assignment which assigns a proportion of the trips to each of the acceptable routes is referred to as multipath proportional assignment.

10. In the larger transportation studies the assignment of trips to the public transport network is simpler than the assignment of trips to the highway network because the routes taken by public transport facilities are closely defined.

Solutions

1. Correct.

2. Incorrect. A choice trip is one in which the tripmaker has a choice of mode. In most surveys the choice of mode is between the private car and public transport. All public trips are assigned to the public transport network but some of the private car trips are assigned to the highway network and others are restrained and transferred to the public transport networks.

3. Incorrect. Because of the lack of highway capacity in the future it will not be possible to make every trip by the desired mode when that mode is 'drive own car'. Not all these trips will be made by the public transport network since some of these trips will not be made at all if they cannot be made by car.

4. Incorrect. As trips are assigned to the public transport network the flows on the network begin to approach capacity and the travel times are increased. This will require a reappraisal of the trip distribution process.

5. Correct.

6. Correct.

7. Incorrect. During the traffic assignment process the computer calculates the least cost travel paths between all the origins and destinations. These routes are referred to as the minimum tree network and usually a selection of these routes are printed out so as to allow a check to be made on the accuracy of the assignment process.

8. Incorrect. Capacity assignment is an assignment process in which allowance is made for the increase in journey cost caused by an increase in flow on the road network This prevents the unrealistic situation developing in which a considerable number of trips may be assigned to a given route because the cost of using this route is marginally less than an alternative route to which only a small number of trips are assigned.

9. Correct.

10. Incorrect. In a large study the assignment of public transport trips is normally more complex because it is necessary to take into account walking and waiting times and also the delay caused by the transfer from one route to another.

10

The evaluation of transportation proposals

The object of the simulation of the land-use/transportation process is to estimate the trips that will be attracted to a proposed future transportation system for a given pattern of land-use development. To compare effectively different transportation proposals it is imperative that each proposal should be evaluated.

There are several grounds on which the end product of the simulation process must be judged. These are tabulated below:

1. It is necessary to be certain that the numerical results output by the computer are realistic and all the computer programs are functioning correctly.
2. It is necessary to check that the predicted future transportation requirements can be met by the proposed transportation system being evaluated.
3. It is necessary to consider the environmental effects of the operation of the proposed transportation system.
4. It is necessary to estimate the economic consequences of the provision and operation of the system under evaluation.

These sections will now be considered in greater detail.

Program checking to ensure that numerically correct results in accordance with the mathematical model are incorporated into the program is a very tedious process and even after experience with transportation programs doubt can sometimes be felt about the validity of the results obtained. A frequent source of error is incorrect input information and this should be carefully checked to ensure that it has been correctly coded and punched.

The second stage in the evaluation of a transportation plan is to ensure that the proposed system is capable of dealing with the flows assigned to it by the simulation process. Many transportation simulation programs deal with 24 hour flows while critical flows usually occur in practice during the peak hours. It may be possible to estimate the peak hour flows from the 24 hour flows or alternatively the transportation programs may be run to deal with a peak hour flow.

Where highway networks are being evaluated it may be found that where there is no restriction on modal choice the highway system may be grossly overloaded. This problem may be overcome by the use of a modal choice relationship that incorporates

71

a travel time by public or private transport ratio or by the arbitrary transfer of trips from the private transport to the public transport system.

Finally it should be remembered that the predictions of future trips on the transport system are dependent on the initial assumptions made. Where all-or-nothing assignment is used, for example, the network may have a section of highway where heavy traffic flows are predicted while an adjacent link, which has a longer travel time, may be lightly loaded. In practice the trips would be distributed between the adjacent links and predicted flows should be considered on the basis of corridors of movement.

Until comparatively recently the environmental effects of proposed transport systems have received less consideration than the effects of capacity or economic considerations. It is unfair to criticise transportation planners for ignoring environmental considerations in the past. Normally they have been required to provide the least-cost solution and since damage to the environment has in the past not been given a high economic price, solutions that have damaged the environment have been preferred.

There are however considerable problems in placing a value on environmental factors and this has been partially responsible for the failure to consider them. The most important environmental effect at the present time is that of noise and while it is possible to predict the noise levels that will be caused by major highway proposals, the valuation of the effects of noise is less easy to make.

Transport facilities usually have an adverse effect on the environment: they cause noise and vibration, highway traffic pollutes the air and they frequently intrude both physically and visually. The disturbance to the environment must however be balanced against the result of doing nothing. A situation in which traffic is congested or there is a high accident risk may damage the environment to a greater extent than a new highway or other transport system.

The cost of reducing the damage to the environment caused by a new transport system may however be better spent providing alternative facilities that may improve the environment. This factor should be seriously considered at a time when any new transport proposal arouses intense opposition from those immediately affected.

Economic evaluation of construction costs

It is however the economic consequences of carrying out transportation schemes which have received increasing attention during the last decade and it is becoming increasingly necessary to justify any new proposal on economic grounds.

This evaluation of a transportation scheme on economic grounds is referred to as a cost–benefit analysis. As would be expected the costs of carrying out a scheme are compared with the benefits that may be anticipated when the scheme is completed and in operation.

The cost of carrying out a scheme may be conveniently classified as construction costs, land purchase costs and the temporarily increased travel costs caused by construction work. The benefits that accrue from the construction of the scheme are those associated with decreased travel and increased convenience by the users of the new system and also, where public transport systems are being considered, the benefits obtained by the operators of the system. When a new highway system is being evaluated then it is to be expected that there will be a saving in accident costs and these are part of the benefits. There will also be changes in maintenance costs, which in the case of a transport system in which equipment is a considerable part of the capital cost, will result in lower maintenance costs if old equipment is replaced. On the other hand a new highway system will normally increase the area of carriageway and hence the cost of maintenance.

Construction costs can be obtained by normal estimating methods and since it is usual to compare costs and benefits on a long term basis then the construction costs during each year of the construction period should be estimated as accurately as possible. Where it is feasible to construct the scheme over differing time periods, then the economic consequences of such decisions should be evaluated. The costs of land acquisition, compensation, legal and administrative expenses should also be included and if the land has been purchased previously, then the land should be valued at the price at which it could be sold for alternative purposes.

During the construction work necessary for most large transportation schemes, there is usually considerable disturbance to traffic flow on both the public and the private transport networks. The cost of such delays can usually be calculated on the same basis as the value of the benefits obtained from decreased travel times and this will be discussed subsequently. As well as the relatively easily measured values of time lost due to construction delays there are also less easily valued construction costs caused by noise, dust and disturbance to the visual scene.

Where extensive highway works are being evaluated some allowance must be made for the increased cost of maintenance necessitated by the new highways. If discounted calculations are being made then future maintenance costs can be related to the years in which they occur, while if the simplest economic evaluation is required, a fixed annual charge for all types of maintenance is adequate.

Benefits from the provision of facilities

The benefits that are obtained by the construction of a new transport system are considered firstly from the point of view of those benefits that accrue to individual users of the system. For most users of the new system there will be a decrease in the cost of travel and a decrease in the time of travel compared with the situation in which the new system is not introduced.

When the transportation system being evaluated is not likely greatly to influence the pattern of travel, then it is approximately correct to assume that the benefits to the individual users can be obtained by calculating the reduced cost of carrying a fixed number and pattern of trips.

Where however the new transportation system is likely to change the whole pattern of tripmaking, then it is no longer possible to compare the cost of tripmaking with and without the change of system. This point is illustrated by the provision of cheaper and more convenient transport facilities which encourage people to spend a greater proportion of their income on travel.

The London Transportation Study used an approach to this problem in which estimates were made of the traffic that would flow between given points if the new system were or were not introduced. The mean of these two values is then multiplied by the difference in the cost of travel with and without the transport system, and the resulting values are summed over all the possible modes of travel to give the benefit to users of the system.

Benefits to operators of transport systems from either the introduction or the alteration of a transport system can be calculated simply as the change in net receipts after the alteration of the system or the net receipts of a new system.

A controversial aspect of a new or amended transport system is that benefits may be said to accrue to the Central Government from increased road-user taxation. The attitude of government departments in the UK is that all costs on which savings are based are net of indirect taxation. The stated reason for this is that taxation merely represents a redistribution of wealth between tripmakers and the Central Government. This is true of existing highway traffic, which benefits from an improvement in the highway system, but cannot be said to be true when generated traffic is considered.

Savings in accident costs have on the other hand a clearly defined benefit to be included in the evaluation. This is particularly true of highway schemes where improvements and new construction can be expected to produce substantial savings in accidents. There is a difficulty in calculating accident costs because not all highway accidents are reported. Damage-only accidents need not necessarily be reported to the police and the only complete accident record is that of personal injury accidents. For this reason a given type of highway can be expected to have a statistically determined personal injury accident rate. The saving in personal injury accident cost can then be calculated from the difference in personal injury accidents between the 'do nothing' situation and the situation in which the improvement is introduced. The cost of a personal injury accident must therefore include an allowance for the non-injury accidents that will also occur, and must also make some allowance for suffering and grief. The value placed on these factors is therefore somewhat arbitrary.

Relationship of costs to benefits

Once all the costs and benefits of the system being evaluated have been determined, it is necessary to relate costs to benefits in some meaningful way.

The simplest way in which these costs may be related is by means of the first year rate of return in which the net benefits in the first year of operation are expressed as a percentage of the initial capital cost. Where alternative schemes are being compared then the first-year rate of return will indicate which scheme is to be preferred on financial grounds provided the alternative schemes will all take the same time to bring into operation, have the same life and attract trips at the same growth rate during their life.

With most transport systems however these conditions will not be true for the capital cost of the scheme will be incurred during a few years at the inception of the scheme while the benefits will be obtained during the whole life of the scheme. With highway schemes particularly past experience indicates that these benefits will increase with time as traffic growth takes place and it is unlikely that the time scale will remain the same for all the alternative schemes being considered.

For these reasons it is particularly important to make comparisons on a long term basis and the two methods that are most likely to be used for this form of evaluation are either the internal rate of return method or the present value method.

In the present value method the net annual benefits (that is, benefits minus costs) are discounted over the life of the scheme, usually to either the present date or the date when construction will commence. Alternatively the internal rate of return may be calculated and this is the discount rate which makes the present value of the investment zero.

Using both these methods it is possible to include in any year the period maintenance costs that are incurred with any transport system. It is thus possible to evaluate the economic consequences of using short or long life materials and of construction in stages.

Probably one of the most complex systems to evaluate economically is a new highway network because of the difficulty of assessing the savings that occur when the new highways have been constructed in comparison with the situation that would exist if highway traffic increased and had to pass through the existing network.

To determine these cost changes it is necessary to have a valid relationship that connects speed and flow on the highway being considered, and knowing the change in speed to evaluate the savings in time and vehicle operating costs.

An added difficulty is that highway flows vary throughout the day, week and year and the cost-flow relationship does not vary in a linear manner. This means that average flows cannot be used to calculate average operating costs. Instead the concept of a representative flow has been developed in *Road Research Technical Paper* 75, 'The economic assessment of road improvement schemes' . The representative flow is defined as being greater than the mean hourly flow and less than the maximum hourly flow, and when substituted in a cost-flow relationship it gives the same total cost of travel as when grouped hourly flows are used.

Where the distribution of flows is known, then *Road Research Technical Paper* 75 shows that the representative flow may be obtained from

$$r = \sum_{i=1}^{p} \left(\frac{n_i q_i}{1 - bq_i} \right) \bigg/ \sum_{i=1}^{p} \left(\frac{n_i q_i}{1 - bq} \right)$$

where r = representative flow,

p = number of groups into which hourly flows are grouped,

n_i = number of hours when flow is in group i,

q_i = hourly flow in group i (i = 1, 2, ... p),

q = mean hourly flow over a year,

b = a constant in the equation connecting average speed v_i to the hourly flow q_i in the expression $v_i = a(1 - bq_i)$ from which it follows that the maximum capacity of the highway is at most $1/b$.

Approximations to the representative flow may be derived. Firstly if the speed in the peak period is above approximately 35 km/h when the flows are small compared with the capacity of the road

$$r = \bar{q}(1 + u^2) \tag{10.1}$$

where $\bar{q}$ is the annual mean hourly flow,

u is the coefficient of variation of hourly flows (the standard deviation divided by the mean).

Secondly if the speed in the peak period is between approximately 18 and 35 km/h and the coefficient of variation of hourly flows is less than 1·0, then

$$r = \bar{q} \left[1 + \frac{u^2}{(1 - b\bar{q})} \right] \tag{10.2}$$

Typical values of b are given below in table 10.1

TABLE 10.1

Road width		Central urban roads	Faster urban roads	Rural roads (single carriageway)
m	ft			
4·8	16	0·000945	0·000592	0·000264
6·0	20	0·000699	0·000424	0·000197
7·2	24	0·000555	0·000330	0·000157
9·9	33	0·000379	0·000220	0·000108
14·4	48	0·000248	0·000142	0·000071

For rural areas the coefficient of variation of the average distribution of hourly flows may be taken as 0·73 giving a value of $r = 1·53\bar{q}$.

When information relating to August flows is available, the representative flow may be calculated from

$$r = A \left(0·1076H + 0·0260 + \frac{1·60}{150 + 874H} \right) \tag{10.3}$$

where A = the flow in an average 16-hour August day,

H = the proportion of August traffic accounted for by medium and heavy goods vehicles.

The relationship between flow, speed and travel cost for a highway

Once the representative flow has been calculated it is necessary to be able to estimate the average speed of vehicles at this flow. An interesting research project on the relationship between speed and flow on suburban main roads has recently been carried out by Freeman, Fox and Associates in co-operation with the Transport and Road Research Laboratory[2]. They carried out observations at several sites on main roads in suburban areas and found that the differences in the speed/flow relationships obtained at the sites could be attributed to differences in the physical conditions prevailing at the sites. The most important of these were:

1. The number of intersections.
2. The proportion of dual carriageway.
3. The amount of access available to the adjacent land uses.
4. The amount of development on the adjacent land.

The relationships obtained apply only to routes on the outskirts of large conurbations which carry a high proportion of through traffic and which are normally classified as principal, trunk or A routes. The speeds given by the relationships are journey speeds, which include stopped time at intersections along the route.

The speed/flow relationship is assumed to be linear and of the form

$$V = V_0 + S \left(\frac{Q - 300}{1000} \right) \tag{10.4}$$

where V is the journey speed measured in kilometres/hour,

Q is the one-way flow of all vehicles, excluding solo motor cycles and pedal cycles, expressed in vehicles per hour per standard lane (a standard lane width is defined as 3·65 metres. In the case of a single carriageway, Q is found by dividing the one-way flow by half the carriageway width expressed in standard lanes. In the case of a dual carriageway, Q is found by dividing the one-way flow by the width (expressed in standard lanes) of the carriageway on which the flow is travelling),

V_0 is the 'free speed' which is the speed of the traffic stream when Q is 300 veh/h per lane,

S is the slope, which is the change in speed observed for an increase in the value of Q of 1000 veh/h per lane.

Whenever possible the report recommended that a value of free speed is obtained by direct observation of the traffic stream at flows of 300 veh/h per lane, or thereabouts, since use of an observed value takes into account local effects which cannot otherwise be incorporated in a general formula.

Where direct observation is impossible, a value of free speed may be obtained from the following formula

$$V_0 = \left(50 + \frac{d}{10}\right) - 10(i - 0·8) - 0·15(a - 27·5) \qquad (10.5)$$

where V_0 is the free speed, measured in km/h, at a composition of 85 per cent light vehicles, 13·5 per cent heavy vehicles and 1·5 per cent buses (light vehicles are defined as vehicles with three or four tyres, heavy vehicles as vehicles with more than four tyres, and buses as buses and coaches on scheduled, local, stopping services),

i is the density of major intersections measured in intersections/km (a major intersection is defined as an intersection such as a roundabout or traffic signals where the road under consideration does not have priority).

a is the accessibility (accessways/km) (accessibility is defined as the sum of the density of minor intersections and the density of private drives. A minor intersection is a point where the road under consideration meets one or more public roads over which it has priority. A private drive is a vehicular accessway other than a public right of way: when counting private drives, the total for both sides of the road should be used (even for dual carriageways),

d is the proportion of dual carriageway, measured as a percentage.

If values of accessibility are not available, typical values are: high: 70/km, medium: 27·5/km, low: 10·0/km.

High values are typical of areas of dense housing or commercial activity, such as shops and service stations where the development is continuous along the length of the road; medium values apply to roads through areas of mixed development, including some open space, and low values correspond to areas only partially developed, or with restricted direct access and served by service roads.

The slope S of the equation is given by

$$S = -25 - 1·33[V_0 - (50 + 0·1d)] - 30(i - 0·8) - 0·4(b - 65) \qquad (10.6)$$

where b is the proportion of the roadside that is developed (taking the average of both sides of the road), measured as a percentage.

If a value of free speed is determined by direct observation, the value actually observed should be corrected to the standard composition of traffic, before inserting it in the equation for the slope. This correction is given by

$$\frac{\text{standard } V_0}{\text{actual } V_0} = \frac{100}{102 - 0\cdot133(P_H + P_B)} \tag{10.7}$$

where P_H is the percentage of heavy vehicles in the stream observed,

P_B is the percentage of buses in the stream observed.

Should the value of free speed be determined by direct observation, it should be corrected for the effect of weather and road conditions at the time of the observations before deducing the slope. The correction, which should be added to the observed free speed, is given by

$$0\cdot84w$$

where w is a composite factor taking into account the effects of visibility, weather and road surface. The appropriate value of w is found by adding the three numerical values assigned to conditions of visibility, weather and road surface shown in table 10.2. As these corrections will normally be small, they will only be of significance when observations are made under extreme conditions.

TABLE 10.2

Visibility	Value	Weather	Value	Road surface	Value
light	0	dry	0	dry	0
dark	1	rain	1	wet	1
twilight	2	falling snow	2	slush	2
		mist	4	snow	3
		fog	5	ice	4

The free speed V_0 obtained using equation 10.5 applies to a standard composition of traffic. If it is required to predict a free speed at some other composition of traffic, the correction given in equation 10.7 should be applied. This non-standard value of the free

TABLE 10.3

Variable	Lower limit	Upper limit
V_0	38	71
S	75	0
i	0	2
a	5	75
d	0	100
b	20	100
P_H	5	40
P_B	0	7
w	0	7

speed is used to fix the vertical position of the speed/flow line. The standard value of V_0 should be used when deducing the slope.

It is important to remember that the variables used are subject to upper and lower limits: these limits are shown in table 10.3.

Once representative average speeds have been derived for the highways under consideration, it is necessary to relate these changes in speed to changes in operating costs on the highways. Fuel, tyre wear, maintenance and depreciation costs vary with speed and relationships between speed and running costs have been given in various Departmental publications.

The Department of the Environment issued values of running costs in 1971 to be used in the assessment of highway schemes. An extract from these values is given in table 10.4.

TABLE 10.4

Vehicle speed (km/h)	New pence/ km	Vehicle speed (km/h)	New pence/ km
17	6·78	54	2·88
18	6·45	55	2·85
19	6·16	56	2·83
20	5·91	57	2·80
21	5·68	58	2·78
22	5·47	59	2·75
23	5·28	60	2·73
24	5·11	61	2·70
25	4·96	62	2·68
26	4·82	63	2·65
27	4·69	64	2·63
28	4·56	65	2·60
29	4·44	66	2·58
30	4·32	67	2·56
31	4·21	68	2·54
32	4·11	69	2·52
33	4·01	70	2·51
34	3·93	71	2·49
35	3·85	72	2·47
36	3·78	73	2·46
37	3·70	74	2·44
38	3·63	75	2·42
39	3·56	76	2·41
40	3·50	77	2·39
41	3·44	78	2·37
42	3·38	79	2·36
43	3·32	80	2·34
44	3·26	81	2·33
45	3·20	82	2·31
46	3·15	83	2·30
47	3·10	84	2·29
48	3·06	85	2·28
49	3·02	86	2·27
50	2·99	87	2·26
51	2·96	88	2·25
52	2·93	89	2·25
53	2·90	90	2·24

Example on the first year rate of return

The application of these techniques to the evaluation of highway proposals is illustrated in the following example. It is estimated that in the design year the 16 h August traffic flows given below will occur from zone i to zone j on the highway links 1, 2 and 3. A new 2-lane link between i and j, numbered 4, is to be constructed with a length of 18·5 km and the travel speed on the new highway is estimated to be 100 km/h.

On the existing routes the average speed of vehicles is measured when the volumes flowing and the roadway and traffic conditions are as given in table 10.5.

TABLE 10.5

1	2	3	4	5	6	7	8	9	10	11	12	13
Link	Travel time (min)	Distance (km)	Volume vehicles/lane	Design year 16 h August flow on link (vehicles)	Visibility	Weather	Road surface	d	i	b	Proportion of heavy goods	Proportion of buses
1	18·5	13·9	305	6900	light	rain	wet	100%	2	100%	15	5
2	17·0	14·5	289	5100	light	dry	dry	100%	3	100%	12	7
3	20·5	15·6	310	6200	light	mist	wet	100%	3	100%	16	5

Calculate the benefits of constructing this additional highway link in the design year.

The first stage in the evaluation of any benefits that will accrue from the construction of the new highway link is to calculate the cost of travel from zone i to zone j if the new link is not constructed. The cost of travel will depend upon the speed on the link and the speed will be estimated by means of a speed/flow relationship using as the flow the representative flow.

Using the relationship given previously in equation 10.3

$$r = A \left(0·1076H + 0·0260 + \frac{1·60}{150 + 874H} \right)$$

where A = flow in an average 16 h August day,

H = the proportion of medium and heavy goods vehicles in the August flow.

TABLE 10.6

Column	14
Link	*r* (veh/hour)
1	330
2	230
3	339

It is now necessary to calculate the speed on the highway links at the representative flows. This may be done using the speed/flow relationship given previously in equation 10.4.

$$V = V_0 + S \left(\frac{Q - 300}{1000} \right)$$

where V_0 is the 'free speed', which is the observed speed at approximately 300 vehicles/ lane per hour. The observed speed must however be first corrected for traffic conditions to the standard composition of traffic using the relationship previously given in equation 10.7

$$\frac{\text{standard } V_0}{\text{actual } V_0} = \frac{100}{102 - 0 \cdot 13(P_H + P_B)}$$

TABLE 10.7

Column	15	16	17	18
Link	Actual V_0	P_H	P_B	Standard V_0
1	45·1	0·15	0·05	44·2
2	51·2	0·12	0·07	50·2
3	45·7	0·16	0·05	46·0

A further correction has to be made for roadway conditions, given by $0 \cdot 84w$.

TABLE 10.8

Column	19	20	21	22	23	24
Link	Visibility correction	Weather correction	Road surface correction	Total w	Correction	Corrected V_0
1	0	1	1	2	1·7	45·9
2	0	0	0	0	0	50·2
3	0	4	1	5	4·2	50·2

To complete the speed/flow relationship the slope S in equation 10.6 has now to be calculated. It is given by

$$S = -25 - 1 \cdot 33 V_0 - (50 + 0 \cdot 1d) - 30(i - 0 \cdot 8) - 0 \cdot 4(b - 65)$$

where i is the density of major intersections measured in intersections/km,

d is the proportion of dual carriageway measured as a percentage,

b is the proportion of roadside that is developed measured as a percentage.

These values are known for the links 1 to 3 and are tabulated in columns 25 to 29 (table 10.9) together with the resulting value of S. It should be noted that the value of S is outside the limits given in table 10.3 and the speed/flow line should be used with caution.

TABLE 10.9

Column	25	26	27	28	29
Link	V_0	i	d	b	S
1	45·9	2	100	100	−93·75
2	50·2	3	100	100	−141·75
3	50·2	3	100	100	−141·75

It is now possible to calculate the speed on the links when the flow is the representative flow using the speed/flow relationship. These values are given in columns 30 to 31 together with the cost of using the link which is given in column 32 (table 10.10).

TABLE 10.10

Column	30	31	32
Link	Representative flow (veh/lane hour)	Speed (km/h)	Average vehicle running cost (p/km)*
1	330	43·1	3·31
2	230	52·5	2·92
3	339	42·3	3·36

* Average vehicle running cost estimated from Table 10.4

To obtain the annual flow from the 16 h August flow, it is usual in interurban road schemes to use the relationship

$$\text{annual flow} = 300 \times 16 \text{ h August flow}$$

(when the percentage of heavy goods vehicles H is 15–20 per cent), or

$$\text{annual flow} = (150 + 8H) \times 16 \text{ h August flow}$$

(when the percentage of heavy goods vehicles H is not 15–20 per cent).
Using these relationships the annual flows are given in column 34 (table 10.11).

TABLE 10.11

Column	33	34
Link	H	Annual flow
1	15%	1 863 000
2	12%	1 254 600
3	16%	1 723 600

The annual cost of operating the three highway links can then be calculated from the sum of the lengths of the links multiplied by the annual flow and the cost of vehicle operation. This calculation is tabulated in columns 35 to 38 inclusive (table 10.12).

TABLE 10.12

Column	35	36	37	38
Link	Length (km)	Annual flow (vehicles)	Cost of operation (p/km)	Total cost of operating link
1	13·9	1 863 000	3·31	£857 150
2	14·5	1 254 600	2·92	£531 200
3	15·6	1 723 600	3·36	£892 450

Σ £2 280 800

To estimate the saving in travel cost caused by the construction of the new link, it is necessary to calculate the travel cost with the new road in operation. Before this can be done, it is necessary to apportion the trips between the four links. This will be carried out on the basis of time and distance and as an approximation the travel times on links 1–3 which exist before the new link is opened to traffic will be used although the diversion of some traffic to the new route where travel speeds are higher will raise the speed on the existing links.

The assignment will be made on the basis of the relationship

$$\text{percentage diverted to new link} = 50 + \frac{50(0{\cdot}62d + \frac{1}{2}t)}{\sqrt{[0{\cdot}62d - \frac{1}{2}t)^2 + 4{\cdot}5]}} \qquad (10.8)$$

where d = distance saving using new link (km),
t = time saving using new link (m)

TABLE 10.13

Column	39	40	41	42	43
Link	Time saving using link 4 (min)	Distance saving using link 4 (km)	Per cent of existing trips diverted	Per cent of existing trips on this route	Per cent of all trips diverted to new route
1	7·4	−4·6	44·2	37·9	16·5
2	5·9	−4·0	55·0	28·0	15·4
3	9·4	−2·9	61·2	34·1	20·9

Σ 52·8

Column 39 was calculated using a travel speed of 100 km/h on link 4;
Column 40 was calculated using 18·5 km as the length of link 4;
Column 41 was calculated from equation 10.7;
Column 42 was calculated from the 16 h August flow on the link;
Column 43 was calculated from column 41 × column 42.

It is now possible to determine the number of trips on all the four links in the design year when the total number of trips between the two zones is 18 200 (total given in column 5).

TABLE 10.14

Column	44	45
Link	Design year trips 16 h August day (vehicles)	Annual flow in design year (vehicles)
1	3895	1 036 070
2	2293	609 938
3	2402	638 932
4	9610	2 556 260

Column 44 was calculated from the difference between columns 4 and 5 multiplied by 18 200 trips.

Column 45 was calculated, in the absence of a precise knowledge of the heavy vehicle content, from the total annual flow of 4 841 200 vehicles given in column 36 divided in the proportion of the 16 h August day trips given in column 44.

To calculate the cost of using these links it is now necessary to calculate the speed on the links at the representative flow. The representative flow is given by equation 10.3 and depends upon the August flow given in column 44 and also on the medium and heavy goods vehicle content of the flow. With the detail given it is difficult to predict the medium and heavy goods vehicle content of the flow on the various links in the design year and so it will be assumed that the medium and heavy goods vehicle content of the flow is the mean of the values given in column 12, that is 14 per cent. The representative flow is then as given in column 46 (table 10.15). There is no necessity to calculate the representative flow on link 4 because the average vehicle speed is assumed to be 100 km/h.

As for the first case when only three links were in operation, it is now possible to calculate the speed at the representative flow using the speed/flow relationships derived previously. These values are tabulated in column 47 (table 10.16)

The total annual cost of operation of the four highway links in the design year may now be obtained from the sum of the products of link length, the total annual volume and the cost of operation per vehicle km. The annual flow is given in column 49 (column

TABLE 10.15

Column	46
Link	r (veh/hour)
1	183
2	108
3	113

TABLE 10.16

Column	46	47	48
Link	r	Speed (km/h)	Cost of operation (p/km)
1	183	56·9	2·80
2	108	77·4	2·38
3	113	76·7	2·40
4	–	100·0	2·14

45), link length in column 50 (column 35), and the cost of operation per vehicle km is given in column 51 (column 48). The cost of using the separate links is tabulated in column 51 (see table 10.17)

TABLE 10.17

Column	49	50	51	52
Link	Annual flow in design year (vehicles)	Link length (km)	Cost of operation (p/km)	Total cost of operating link (£)
1	1 036 070	13·9	2·80	403 250
2	609 938	14·5	2·38	210 500
3	638 932	15·6	2·40	239 200
4	2 556 260	18·5	2·14	1 012 000

Σ £1 864 950

The cost of operation of the highway system in the design year with the new link constructed is the sum of column 52, namely £1 864 950. The cost of operation of the highway system in the design year without the construction of the link is the sum of column 38, namely £2 280 800. The difference between these operating costs, £415 850, is the annual saving in operating costs made possible by the new highway link. There will also be savings in accident costs due to a reduction in highway congestion, but an increase in maintenance costs caused by the increased length of highway in operation.

These costs and savings can be related to the construction cost using the first year rate of return. This can be expressed as

$$\text{first year rate of return} = \frac{£415\,850 + \begin{array}{c}\text{saving in}\\ \text{accident}\\ \text{costs}\end{array} - \begin{array}{c}\text{increased}\\ \text{maintenance}\\ \text{costs}\end{array}}{\text{capital cost of scheme}} \times 100 \text{ per cent}$$

An example of the return from transport investments over a period of time

As has previously been stated the first year rate of return is only an indication of the relative value of a highway or transport investment measured by the financial return in the first year of operation. Where it is required to compare investment in schemes it is usually necessary to compare costs and benefits over a period of time.

The following example illustrates the approach in this type of calculation.

Determine the economic return over a time period of 10 years from the investment of £15 000 in the installation of traffic signals at a highway junction where priority control is in operation.

It can be assumed that delay to vehicles when priority control is in operation can be related to the representative traffic flow by the following expression

$$\text{Delay/vehicle} = 12 + 0.167 \, (\text{traffic volume})^3 \times 10^{-7} \text{ second}$$

It can also be assumed that delay to vehicles when traffic signal control is in operation can be related to the representative traffic flow by the following expression

$$\text{Delay/vehicle} = 20 + (\text{traffic volume})^2 \times 10^{-5} \text{ second}$$

The cost of delay may be taken as £1 per vehicle hour and the additional cost of the maintenance of the signals may be taken as £2000 per annum with increased maintenance costs of £4000 in the 4th and 8th year of operation.

The expected traffic volumes during the next 10 years are given in column 52 (table 10.18) and it can be assumed that for the anticipated traffic conditions the annual flow is 1500 times the representative flow.

TABLE 10.18

51	52
Year of operation	Representative traffic flow
1	1000
2	1100
3	1200
4	1300
5	1400
6	1500
7	1700
8	1900
9	2100
10	2300

The first step in the estimation of benefits is to calculate the delay to vehicles with and without the installation of traffic signal control using the delay/flow relationships given: these values are tabulated in columns 54 and 55 (table 10.19).

TABLE 10.19

Column	53	54	55
		Delay/vehicle (second)	
Year of operation	Representative traffic flow	Priority control	Traffic signals
1	1000	28·7	30·0
2	1100	34·2	32·1
3	1200	41·5	34·4
4	1300	48·7	36·9
5	1400	57·8	39·6
6	1500	68·4	42·5
7	1700	80·6	48·9
8	1900	126·5	56·1
9	2100	166·7	64·1
10	2300	215·2	72·9

The annual reduction in delay caused by the installation of traffic signals is given by the difference between columns 54 and 55 multiplied by fifteen hundred times the representative flow. (The factor of 1500 is derived from a factor of 5 to convert to daily flow and a factor of 300 to convert to yearly flow.) This reduction in delay is valued in

TABLE 10.20

Column	56	57	58	59	60
Year of operation	Reduction in delay/vehicle (second)	Value of annual reduction in delay (£)	Capital and maintenance cost (£)	Net benefit (£)	Discounted benefit (£)
1	−1·3	−541	17 000	−17 541	−16 414
2	2·1	962	2 000	−1 038	−907
3	7·1	3 050	2 000	+1 050	+857
4	11·8	6 392	4 000	+2 392	+1 825
5	18·2	10 616	2 000	+8 616	+6 143
6	25·9	16 188	2 000	+14 188	+9 455
7	31·7	22 454	2 000	+20 454	+12 739
8	70·4	55 733	4 000	+51 733	+30 159
9	102·6	89 775	2 000	+87 775	+47 750
10	142·3	136 370	2 000	+134 370	+68 314

Σ 159 891

column 57 at the rate of £1 per hour of delay to give the economic benefit in each year of operation. Capital and maintenance costs are given in column 58 and the net benefit is then discounted to its present value at a discount rate of 7 per cent in column 60 (table 10.20). The net present value of the scheme is £159 891, and the scheme is thus economically worthwhile.

The COBA program

During the early 1970s the Department of Transport introduced a computer cost-benefit analysis program. The COBA program[3] compares the cost of improving or constructing a road with the benefits obtained by users of the road, and expresses the results of the computation in a monetary term, the net present value. Since the introduction of the program it has been revised in line with increased research.

The COBA evaluation system calculates the cost to road users of the existing highway network and also the cost to road users of the proposed improved network, both discounted over a period of 30 years. The difference between the two user costs is the user benefit. If the construction and maintenance costs of the proposed improvement are also discounted over 30 years then the difference between discounted user benefit and discounted construction cost is the net present value of the scheme.

The COBA program calculates changes in travel time costs, vehicle-operating costs and accident costs as they arise over a modelled road network, each component of which is carrying specified traffic flows. The travel cost for each link and junction are summed to yield the total travel costs over the network. The calculations are made with and without the road scheme under evaluation.

In the COBA calculations the trip matrix is assumed to be fixed and a comparison is made using this matrix of flows between the total costs before and after the proposed improvement. The consequence of using a fixed trip matrix is that only re-assignment of trips that is caused by the proposed improvement are considered. Redistribution of trips that is due to changes in trip destination as a result of the proposed improvement under evaluation is not considered in the COBA program. Similarly, the generation of trips as a result of the improvements in the system, the change in the mode of transport and the changes in the time of day when the trip is made are ignored in the program. The Department of Transport believes that these are realistic assumptions and states that there is very little evidence that in most road schemes the other effects which are ignored are significant. Circumstances in which these effects may be important are given as: in congested urban areas where there is likely to be restraint on car trips; on long inter-urban schemes where significant modal changes may be expected; in major estuary crossings. It is stated that, in most cases, the variable trip matrix evaluation of benefits is unlikely to yield more than in the region of 10 per cent extra benefit over fixed trip evaluation.

For a detailed explanation of the operation of the program the reader should refer to the COBA manual[3] but an outline of the calculation of construction expenditures and user costs will be useful as preliminary reading.

As a first step in the economic appraisal the alternative options must be defined. The minimum number of options is two: a 'Do-Minimum' and a 'Do-Something' option. The former is usually the existing road network without any improvement;

but if any improvements are to be made regardless of the consequences of the economic appraisal currently being carried out, then the existing road network should be considered to include the improvement. This type of improvement occurs frequently when a junction improvement is to be carried out as part of building development.

Normally there will be several 'Do-Something' options, each of which can be tested against the 'Do-Minimum' situation.

The initial step in calculating user costs is to input into the program for each link the length, location, accident rate, bendiness, hilliness and similar geometric features. For each classified junction the details should include layout, turning traffic proportions and geometric delay. Details of the existing traffic flows either in the form of 1-hour, 12-hour or 16-hour flows for each link are also input and are then converted by the program into total yearly flows using factors that depend on day, month and length of the initial traffic survey.

The program then divides the flow into the different vehicle classes, cars, buses and light goods vehicles, using proportions input by the program user. The next step is to calculate for each link for the differing vehicle classes the yearly flow for a 30-year evaluation period using growth factors taken from either National or other Traffic Forecasts. Finally, these yearly flows are divided by 8760 to give the annual average hourly flow for each vehicle class.

User accident costs are modelled for junctions and links using flows through junctions and along links, and accident rates that depand on junction and link type. Time costs are calculated using delays at junctions and time spent on links, making an allowance for the percentage of in-work and non-work time. Vehicle operating costs are calculated from a knowledge of the speed on the link and from vehicle operating costs for each vehicle type.

These three types of cost are then summed for each year of the evaluation and discounted to the base year of the program for both the 'Do-Minimum' and the 'Do-Something' options, the difference giving the present value of the benefits. Subtraction of the net present value of the construction cost gives the net present value of the scheme.

Valuation of accident savings in COBA

Calculation of the benefits from a reduction in the number and severity of accidents requires the valuation of accidents. The value of a personal injury accident in the COBA program contains three elements:

(a) Direct financial costs to those involved in the accident including damage to vehicles, police and medical costs.
(b) Loss of output of those killed or injured, measured as the present value of loss of earnings and non-wage payments by employers.
(c) An allowance for pain, grief and suffering resulting from personal injury or death.

The average cost per personal injury accident in 1979 in June 1979 prices is given by the Department of Transport as: direct financial costs £1000, lost output £2150 and pain, grief and suffering £1640. It is the costs of personal injury and death that are the main elements in accident valuation. The average value in June 1979 prices are fatality £101 900, serious casualty £4310, slight casualty £100.

Valuation of time savings in COBA

A major benefit of any highway improvement scheme is the time saved by traffic passing along the improved road. It is obviously important to be able to put a value on these time savings so that the net present value of the scheme can be calculated.

Travellers on a highway have a variety of reasons for making the journey. In the COBA program two purposes are recognised: travel in the course of work and travel for other purposes which includes travel to and from work. Time involved in the former is referred to as working time and in the latter as non-working time.

Working time is valued at the cost to the employer of the travelling employee which is the gross wage rate together with National Insurance and pension contributions plus overheads currently valued at 36.5 per cent of the gross wage rate.

Non-working time is valued at a rate that has been derived from traffic studies where travellers were faced with route choices that saved either time or money. From a wide variety of research it has been deduced that on average travellers value their non-working time at approximately 25 per cent of their gross wage rates excluding overheads. Typical values at 1979 prices are: working car drivers 411.2 pence, non-working car drivers 55.4 pence and working bus passenger 272.2 pence.

Valuation of vehicle operating cost savings in COBA

In the COBA program, vehicle operating cost savings comprise fuel, oil, tyres, maintenance and depreciation, items which vary with the use of the vehicle.

The resource cost of fuel consumption is estimated from,

$$C = (a + b/v + cV^2)(1 + mH + nH^2)$$

where C is the cost in pence per kilometre per vehicle,
V is the average link speed in kilometres per hour,
H = average hilliness of link metres per kilometre (total rise and fall per unit distance),
a, b, c, m and n are parameters defined for each vehicle type.

This equation produces a high cost at low speeds (that is, for stop/start motoring), while the adjustment for gradient makes allowances for increased fuel consumption on gradients.

Marginal resource cost of oil and tyres are taken as a fixed sum per kilometre. Maintenance costs are partly assumed to vary with distance travelled and partly with speed.

Depreciation for goods and public service vehicles is assumed to be entirely related to distance travelled. Depreciation for cars is related both to distance travelled and to the passage of time.

Additionally, in calculation of the savings in vehicle operating costs an allowance is made for the fact that increased speeds result in a reduction in the number of commercial vehicles, public service vehicles and working cars that need to be provided.

References

1. Road Research Laboratory, Ministry of Transport, *The economic assessment of road improvement schemes,* HMSO, London (1968)
2. Freeman, Fox and Associates/Road Research Laboratory/Department of the Environment, *Speed flow relationships on suburban main roads,* London (1972)
3. COBA 9 Manual, Department of Transport, London (1981)

PART 2

ANALYSIS AND DESIGN FOR HIGHWAY TRAFFIC

11

The capacity of highways between intersections

The capacity of a highway may be described as its ability to accommodate traffic, but the term has been interpreted in many ways by different authorities. Capacity has been defined as the flow which produces a minimum acceptable journey speed and also as the maximum traffic volume for comfortable free-flow conditions. Both these are practical capacities while the Highway Capacity Manual[1] defines capacity as the maximum number of vehicles which has reasonable expectation of passing over a given section of a lane or roadway in one or both directions during a given time period under prevailing roadway and traffic conditions.

Highway capacity itself is limited by:

1. The physical features of the highway, which do not change unless the geometric design of the highway changes.
2. The traffic conditions, which are determined by the composition of the traffic.
3. The ambient conditions which include visibility, road surface conditions, temperature and wind.

A term that is used to classify the varying conditions of traffic flow that take place on a highway, is level of service. The various levels of service range from the highest level, which is found at a flow where drivers are able to travel at their desired speed with freedom to manoeuvre, to the lowest level of service, which is obtained during congested stop–start conditions.

Level of service

The level of service afforded by a highway to the driver results in flows that may be represented at the highest level by the negative exponential headway distribution when cumulative headways are being considered, by the double exponential distribution as the degree of congestion increases and by the regular distribution in 'nose-to-tail' flow conditions.

To define the term level of service more closely, the Highway Capacity Manual gives six levels of service and defines six corresponding volumes for a number of highway types. These volumes are referred to as service volumes and may be defined as the

95

maximum number of vehicles that can pass over a given section of a lane during a specified time period while operating conditions are maintained corresponding to the selected or specified level of service. Normally the service volume is an hourly volume.

Individual road users have little knowledge of the flow of vehicles along a highway but they are aware of the effect of high volume on their ability to travel at reasonable speed, comfort, convenience, economy and safety. The factors involved in evaluating a level of service may include the following:

1. Speed and travel time based not only on the operating speed but also the overall travel time.
2. Traffic interruptions or restrictions measured by the number of stops/km and the delays involved, and the speed changes necessary to maintain pace in the traffic stream.
3. Freedom to manoeuvre as necessary to maintain the desired operating speed.
4. Safety, accidents and potential accident rates.
5. Driving comfort and convenience as affected by highway and traffic conditions and also the degree to which the service provided by the roadway meets the requirements of the driver.
6. Economy considered from the point of view of vehicle operating costs.

It is difficult to incorporate all these features in a standard level of service and so two have been selected; they are:

1. Travel speed.
2. The ratio of demand volume to capacity, or, the ratio of service volume to capacity. This ratio is often referred to as the v/c ratio.

The travel speed is the operating speed in rural areas or the average overall travel speed (including stops) in urban areas or elsewhere where the flow is interrupted. The operating speed is the highest overall speed at which a driver can travel on a given highway under favourable weather conditions and under prevailing traffic conditions without at any time exceeding the design speed. The Highway Capacity Manual gives values of speed and volume/capacity which define levels of service for each of the following types of facilities:

(a) Freeways and other expressways.
(b) Other multilane highways.
(c) 2 and 3 lane highways.
(d) Urban arterial streets.
(e) Downtown streets (approximate only).
(f) Intersections and weaving sections etc.

The levels of service are:

Level of Service A Free flow, low volumes, high speeds, traffic density low. Speeds controlled by driver desires, speed limits and physical roadway conditions. Drivers are able to maintain their desired speeds with little or no delay.

Level of Service B Stable flow but operating speeds beginning to be restricted by traffic conditions. Drivers still have reasonable freedom to select their speed and lane of operation. There is a reduction in speed but it is not unreasonable.

The lower limit of this level of service (lowest speed–highest volume) has been associated with service volumes used in the design of rural highways.

Level of Service C Still in zone of stable flow but speeds and manoeuvrability are more closely controlled by the higher volumes. Most drivers are restricted in their freedom to select their own speed, change lanes or pass.

Relatively satisfactory operating speeds are possible. This level of service is suitable for urban design.

Level of Service D Approaching unstable flow with tolerable operating speeds being maintained though affected considerably by changes in operating conditions.

Fluctuations in volume and temporary restrictions to flow cause considerable drops in operating speed. There is little freedom to manoeuvre and comfort and convenience are low, but this level of service may be tolerated for short periods.

Level of Service E Flow is unstable at this level of service with stop–start flow: speeds seldom exceed 50 km/h. Demand is at or near the capacity of the highway.

Level of Service F Forced flow takes place at this level of service, speeds are low and volumes are below capacity. This level is found in queues which are backing up. At the lowest level the traffic stops.

The levels of service may be represented on the operating speed, volume/capacity curve illustrated in figure 11.1.

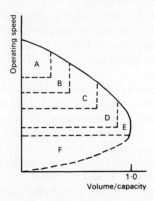

Figure 11.1 The relationship between level of service and the operating speed, volume/capacity envelope

Capacity and service volumes

The Highway Capacity Manual gives service volumes for many highway types and in Table 11.1 are reproduced the values of service volumes for a dual carriageway, restricted access highway with 3 lanes in each direction.

TABLE 11.1 Level of service and maximum service volume for freeways and expressways under uninterrupted flow conditions

Level of service	Description of flow	v/c ratio 6 lane freeway (3 lanes in each direction)	Maximum service volume (passenger cars, one direction)			
A	free flow	≤0·40	2400			
B	stable flow	≤0·58	3500			
		*P.H.F.	0·77	0·83	0·91	1·00
		(peak hours)				
C	stable flow	≤0·80	3700	4000	4350	4800
D	approximately unstable flow	≤0·90	4150	4500	4900	5400
E	unstable flow	≤1·00	6000 approx.			
F	forced flow		0 → capacity (6000)			

$$* \text{ Peak Hour Factor (P.H.F.)} = \frac{\text{number of vehicles in peak hour}}{12 \times \text{number of vehicles in peak 5 min}}$$

In Great Britain the terms standard and maximum working levels of hourly flow are used to describe highway capacity. Demand is measured in terms of average daily flow and peak hourly flow. The former is the estimated average 16 h daily flow (7-day average 0600–2200 hrs) for the most heavily trafficked month in the design year. The latter is the highest estimated flow for any specific hour of the week averaged over any consecutive 13 weeks during the busiest period of the year in the design year. The average daily flow is used for the primary assessment of rural road schemes, while the peak hourly flow is used in detailed design to ensure that the layout standards are adequate from traffic aspects. For motorways and rural all-purpose roads the design flows were first given in the Department of the Environment Technical Memorandum H6/74 and supersede the values given in Layout of Roads in Rural Area[2] while for urban roads it was given in Technical Memorandum H5/75 which supersede the values given in Roads in Urban Areas.[3]

Table 11.2 gives design flows as laid down in Technical Memorandum H6/74. It should be noted that flows are now given in vehicles per hour. This is because passenger car equivalents have been found to be unreliable and the use of vehicles with variants for the proportion of heavy vehicles are more meaningful. A range of flows is given for design purposes and also for assessing the adequacy of existing roads. Memorandum H6/74 points out that with increasing flows above the standard hourly design levels there will be a progressive decrease in vehicle speeds, decline in driver comfort and corresponding increase in congestion and operational costs. These effects will be accelerated once the maximum working flow levels have been passed.

The appropriate allowable peak hour design flow level depends on flow variations throughout the day. The upper values of design flow are only possible when the flow throughout the day is nearly uniform.

These flow variations are measured by the peak hour/daily flow ratio (PDR) which for dual carriageways is the ratio of the hourly peak flow in one direction to the highest monthly average daily flow over 16 h in both directions. For single carriageways the ratio is similar, but with the peak hourly flow measured in both directions.

For urban roads which are designed for peak hour flows, the maximum hourly flow

TABLE 11.2   Design flows for motorways and rural all-purpose roads.
When heavy vehicles comprise ⊁15 per cent of flow

| Road type | Peak hourly flow veh/hour/car'way | | 16 h average daily flow (both directions) | | |
	Standard	Max. working	Min.	Max. within normal PDR range	Absolute max.*
Rural motorways					
dual 2-lane	2400	3200	35 000	45 700–48 000	56 000
dual 3-lane	3600	4800	45 000	68 200–72 000	85 000
dual 4-lane	4800	6400	70 000	91 400–96 000	115 000
All-purpose dual carriageway roads					
Dual 2-lane	2400	3200	17 000	33 700–45 000	45 000
Dual 3-lane	3600	4800	35 000	50 500–60 000	60 000
All-purpose single-carriageway roads					
10 m wide†	2600	3000	20 000	23 100–30 000	30 000‡
10 m wide	1900	2300	12 000	17 700–23 750	25 000
7·3 m wide	1200	1600	2000	12 300–15 000	17 000

* commensurate with exceptionally low values of PDR only
† where centres of interchange are more than 3 km apart add 400 vph
‡ only for grade-separated schemes

TABLE 11.3   Design flows for urban roads when heavy vehicles comprise ⊳ 15 per cent of flow

| Road type | Peak hourly flows veh/hour/both directions | Peak hourly flows veh/hour/carriageway | |
	single 2-lane 7·3 m wide	dual 2-lane	dual 3-lane
Urban motorways		3600	5700
All-purpose roads, no frontage access, no standing vehicles, negligible cross-traffic	2000	3200	4800

level for all-purpose dual carriageway roads are used as their design capacity but a higher peak hour design flow level is acceptable for the high-quality single 2-lane all-purpose road. Table 11.3 gives details of these design flows.

Factors affecting capacity and service volumes

Seldom are all roadway and traffic conditions ideal, and capacity and service volumes require adjustment for these departures from ideal conditions. Adjustment for some effects applies equally to capacity for several levels of service whereas other effects differ depending on the level to which they are to be applied. Restrictive physical features incorporated into the design have an adverse effect on its capacity and service volumes. Such elements are called 'roadway factors', and their effect on capacity will now be considered.

Reduced lane width

The Highway Capacity Manual bases service volumes on ideal highway conditions where the standard lane width is 3·65 m (12 ft). Where the width of a traffic lane is reduced below this standard value then reduced service volumes apply. The percentage reductions in capacity are given in table 11.4.

TABLE 11.4   The effect of a reduction in lane width on capacity

Lane width (m)	Two-lane highways (per cent)	Multilane highways (per cent)
3·65 (12 ft)	100	100
3·50 (11 ft)	88	97
3·00 (10 ft)	81	91
2·75 (9 ft)	76	81

Lateral clearance

It is believed that mountable kerbs and vertical kerbs less than 0·15 m (6 in.) high have insignificant effect on traffic operation but retaining walls, lighting columns etc. closer than 1·83 m (6 ft) from the carriageway edge have an adverse effect.

Both these factors of reduced lane width and obstructions at the side of the carriageway are taken into account in Table 3—20 from Layout of Roads in Rural Areas, which is reproduced in table 11.5.

TABLE 11.5   Capacities of roads restricted in width and/or clearance as percentages of the standard capacities

Type of road	Carriageway width in metres	Obstruction one side of road Distance from carriageway in metres					Obstruction on both sides of road Distance from carriageway in metres				
		0	0·5	1·0	1·5	2·0 or more	0	0·5	1·0	1·5	2·0 or more
Dual	11	94	97	98	99	100	90	95	97	99	100
3-lane	10	91	93	94	95	96	87	91	93	95	96
road	9	85	87	88	88	89	81	84	86	88	89
	8·5	77	78	79	80	81	72	76	78	80	80
Dual	7·3	90	96	98	99	100	81	92	97	99	100
2-lane	6·75	87	93	95	97	97	79	89	94	96	97
road	6·0	82	87	89	90	91	74	84	88	90	91
	5·5	73	78	79	80	81	66	75	78	80	81
2-lane	7·3	85	90	95	98	100	70	79	88	96	100
single	6·75	73	77	81	85	86	60	68	76	83	87
carriage-	6·0	66	69	72	76	77	54	61	68	74	78
way road	5·5	60	63	66	69	70	49	55	62	68	71

Where a shoulder is provided with a width of at least 1·22 m (4 ft) adjacent to a traffic lane with a width less than 3·65 m (12 ft) then the Highway Capacity Manual recommends that the effective width of the lane may be increased by 0·305 m (1 ft).

Alignment

Where the highway alignment is such that speeds below the design speed are necessary or where overtaking is limited by lack of adequate sight distances then there is likely to be a loss of capacity. Layout of Roads in Rural Areas gives the effect of inadequate overtaking sight distance on capacity and this is reproduced as table 11.6.

TABLE 11.6   Effect on design capacity and speed where minimum overtaking sight distances are not provided

Percentage of road length with substandard overtaking sight distance	0	20	40	60	80	100
Percentage of standard design capacity	100	90	80	65	50	30
Estimated reduction in average speed (km/h) of a two-lane road carrying 900 p.c.u./h	0	6	12	18	24	30

The Highway Capacity Manual also gives an indication of this effect. Here the design speed for each horizontal and vertical curve is taken from design tables. Each curve has an area of influence calculated from acceleration and deceleration lengths taken in many cases to be 244 m (800 ft), while tangent sections have a maximum speed of 120 km/h (70 m.p.h.). The average speed on the highway is then calculated and the reduction in capacity due to this speed reduction is estimated from table 11.7.

TABLE 11.7   The reduction in capacity due to reduced speeds

Average highway speed		Capacity as percentage of ideal	
km/h	m.p.h.	Multilane	2-lane highways
120	70	100	100
100	60	100	98
80	50	96	96
60	40		95
50	30		94

Effect of gradient

The effect of gradient on highway capacity can be most conveniently summarised under the following headings.

1. They are often associated with reduced passing sight distances: this effect is taken account of in alignment.
2. Safe headways are less on uphill grades and greater on downhill grades than on horizontal highways.
3. Trucks with normal loads travel more slowly on up-grades but cars negotiate 6–7 per cent grades at speeds above which capacity occurs.

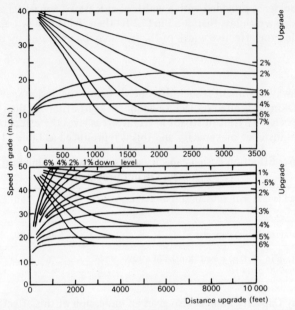

Figure 11.2 Effect of length and steepness of gradient on speed of average trucks on (upper) two-lane and (lower) multilane highways (adapted from ref. 1)

The effect of power/weight ratio on speed up a gradient can be illustrated by figure 11.2, reproduced from the Highway Capacity Manual.

A knowledge of the speed on the grade does not in itself allow a direct calculation of the effect on capacity; this must be calculated from a knowledge of the vehicle composition. Normally it is the average speed over the grade that is important and this is given in detail in the Highway Capacity Manual, an extract from which is given in figure 11.3.

From a knowledge of the average speed on the grade it is possible to calculate the effect of individual vehicle types on the flow by the use of passenger car equivalents.

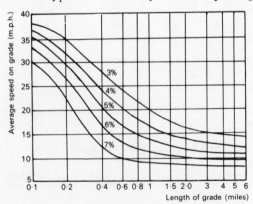

Figure 11.3 Average speed of typical truck over entire length of grade on two-lane highways (adapted from ref. 1)

Traffic composition

In addition to the restrictive effects of roadway factors the composition of the traffic affects the capacity of a highway. Normally the effect of traffic composition is allowed for by the conversion of the flow expressed in vehicles per hour into passenger car units, there being an accepted set of passenger car equivalents for different highway situations.

The effect of gradient may be expressed in terms of passenger car equivalents. For general descriptions of the gradient the p.c. equivalents given in table 11.8 may be used in conjunction with the levels of service used in the Highway Capacity Manual. Where however the effect of slow moving vehicles on single gradients is being investigated then it is possible to arrive at a passenger car equivalent using the graph given as figure 11.4, which was derived for standard trucks and buses in the United States.

TABLE 11.8   Passenger car equivalents for a standard truck under U.S. conditions

Vehicle	Level of service	Level	Rolling	Mountainous
truck	A	3	4	7
	B and C	2·5	5	10
	D and E	2	5	12
bus	all service levels	2	4	6

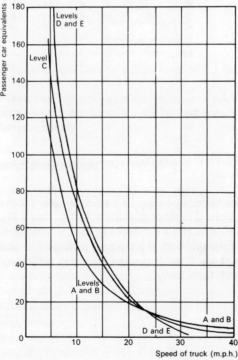

Figure 11.4 Passenger car equivalents for various average truck speeds on two-lane highways (adapted from ref. 1)

An indication of the effect of gradients on traffic flow in Britain can be gained from information given in Department of Transport Technical Memorandum H1/74 which gives design flows on gradients as a percentage of the design flow on a level road. An abstract of this memorandum is given in table 11.9.

TABLE 11.9 Design flow factors for hills expressed as percentages of those on level—single carriageways

Average gradient (per cent)	% heavy vehicles	Length of gradient (km)						
		0.4	0.6	0.8	1.2	1.6	2.4	3.2
3	20	100	78	67	52	44	39	37
	15	100	82	71	57	49	44	42
	10	100	86	77	65	57	52	50
	5	100	92	86	77	71	67	65
4	20	82	60	45	34	30	28	26
	15	85	62	51	39	35	33	31
	10	89	67	59	47	43	40	38
	5	94	83	72	62	58	55	53
5	20	67	47	32	26	24	23	22
	15	71	50	37	31	28	27	25
	10	77	55	44	38	35	33	32
	5	86	69	60	53	50	48	46
6	20	50	33	24	21	20	19	18
	15	55	37	28	25	24	23	22
	10	63	44	35	31	30	29	28
	5	76	59	50	45	44	43	42
7	20	33	24	19	17	17	16	15
	15	37	27	22	20	20	19	18
	10	45	34	29	26	25	24	23
	5	60	48	42	39	38	37	36

Notes: Length of gradient is to be taken from centre of sag vertical curve at foot to centre of hog vertical curve at summit.

In British practice the composition of traffic is largely taken into account by correcting for an abnormal number of heavy vehicles. This is illustrated in tables 11.2 and 11.3 where capacities are reduced for heavy vehicle contents.

The capacities of roundabouts as given by equations 20.1 and 23.1 are also reduced by 4 per cent and 8 per cent when the heavy vehicle content is 15 to 20 and 20 to 25 per cent respectively.

References

1. Highway Capacity Manual. Highway Research Board Special Report 87 (1965)
2. Layout of Roads in Rural Areas. Ministry of Transport, London (1968)
3. Roads in Urban Areas. Ministry of Transport, London (1968)

Problems

1. Are the following statements true or false?

(a) The capacity of a highway must be related to the appropriate traffic flow conditions acceptable to highway users.

(b) The capacity of a highway as defined in the Highway Capacity Manual represents a highly congested level of flow.

(c) At a high level of service it can be expected that drivers will experience considerable difficulty in maintaining their desired speed.

(d) Low volumes of flow are associated with both high and low operating speeds.

(e) Higher traffic capacities are allowed in situations where a large proportion of the peak hour flow is concentrated into a short period of the peak hour.

(f) In rural areas peak hour flows are not of considerable importance and demand is normally expressed in terms of Average Daily Flow.

(g) The capacity of a highway is reduced if the width of the carriageway results in traffic lanes which have a width less than 3·65 m.

(h) The passenger car equivalent of a commercial vehicle on a gradient is greater at low levels of service than at high levels of service.

(i) In Great Britain a high value of PDR indicates high peak flows in relation to average flow.

(j) On British highways a high value of PDR combined with low daily flows indicates long congested peak periods.

Solutions

1.
(a) This statement is correct. In urban areas a greater degree of highway congestion is tolerated and so higher design capacities are used for urban highways than for rural highways.

(b) This statement is correct. The capacity of a highway as defined in the Highway Capacity Manual is the maximum possible traffic volume which the road will carry under ideal highway and traffic conditions. It represents a highly congested level of flow with all the vehicles travelling at approximately the same speed.

(c) This statement is incorrect. At a high level of service drivers are able to travel at their desired speed as the service volume/capacity ratio is low.

(d) This statement is correct. Low volumes of flow are associated with both high speed conditions and high levels of service and also with low speed congested conditions when low levels of service obtain.

(e) This statement is correct. Where a large proportion of the peak hour flow is concentrated into a short period of time then a higher degree of congestion is tolerated because these conditions will not continue for a long period of time.

(f)   This statement is incorrect. Peak hour flows are considered in highway design for rural roads as detailed in Technical Memorandum H6/74.

(g) This statement is correct.

(h) This statement is correct. The effect of a commercial vehicle on a gradient is more marked at low levels of service than at higher levels of service because of the greater difficulty in overtaking slower vehicles at lower levels of service.

(i) This statement is incorrect. The higher the value of PDR, the higher are the peak flows in relation to the average flow, and the shorter is the total period over which these peak flows exist and the lower is the off peak level of flow.

(j) This statement is incorrect. A high PDR with low daily flows indicates short peak periods with little congestion.

12

Headway distributions in highway traffic flow

The concept of level of service in highway traffic flow illustrates the differences in the characteristics of the flow which may be examined by a study of the headways between vehicles. Time headways are the time intervals between the passage of successive vehicles past a point on the highway. Because the inverse of the mean time headway is the rate of flow, headways have been described as the fundamental building blocks of traffic flow. When the traffic flow reaches its maximum value then the time headway reaches its minimum value.

If time headways are observed during any period of time the individual values of time headway vary greatly. The extent of these variations depends largely on the highway and the traffic conditions.

On a lightly trafficked rural motorway, where vehicles can overtake at will, a range of headways will be observed from zero values between overtaking vehicles to the longer headways between widely spaced vehicles.

When flow conditions are observed on more heavily trafficked highways there are fewer opportunities to overtake and fewer more widely spaced vehicles. When overtaking opportunities do not exist there is an absence of very small headways and, under very heavily trafficked conditions, all vehicles are travelling at uniform headways as they follow each other along the carriageway.

There are two general approaches to the method of measurement of headways. They may be measured by a device that registers the successive arrivals of vehicles at a fixed point. Alternatively headways may be recorded by aerial photography which records at one instant of time the distribution of headways between successive vehicles.

By the first method it is the time headway distribution that is obtained and in the second method it is the space headway distribution. Because of the ease of observation it is the time headway distribution that has been extensively researched and reported.

When the arrival of vehicles at a particular point on the highway is described, the distribution may either describe the number of vehicles arriving in a time interval or the time interval between the arrival of successive vehicles. The first type is the counting distribution and the second type is the gap distribution.

As great variability in all types of headway may be recorded, it has been described as the most noticeable characteristic of vehicular traffic. For this reason attempts to understand headways have employed statistical methods and probability theory to find theoretical distributions to represent the observed headway distributions.

Haight, Whisler and Mosher[1] studied the relationship between the counting and the gap distribution but usually it is the gap distribution that is studied. It requires a shorter period of observation to collect data for the investigation of gap distributions than for an investigation into counting distributions. In the study of intersection capacity, gaps in the major road flow are used by minor road vehicles to enter the major road and, once again, it is the gap distribution that is of importance.

One of the earliest headway distributions proposed for vehicular traffic flow was proposed by Kinzer[2] and Adams[3] who suggested that the negative exponential distribution would be a good fit to the cumulative gap distribution. Adams illustrated the validity of the negative exponential distributions by observations of traffic flow in London. When this distribution represents the cumulative headway distribution then arrivals occur at random and the counting distribution may be represented by the Poisson distribution. This type of flow may be found where there are ample opportunities for overtaking, at low volume/capacity ratios.

The negative exponential headway distribution

If the traffic flow is assumed to be random then the probability of exactly n vehicles arriving at a given point on the highway in any t second interval is obtained from the Poisson distribution which states

$$\text{probability } (n \text{ vehicles}) = (qt)^n \exp(-qt)/n! \qquad (12.1)$$

where q is the mean rate of arrival per unit time.

Often this distribution is referred to as the counting distribution, because it refers to the number of vehicles arriving in a given time interval. It is the negative exponential distribution however which is most commonly used when describing headway distributions.

The negative exponential distribution can be obtained from the Poisson distribution if there are no vehicle arrivals in an interval t. In this case there must be a headway greater than or equal to t.

$$\text{probability (headway} > t) = \exp(-qt) \qquad (12.2)$$

The mean rate of arrival q is also the reciprocal of the mean headway which can be computed from the observed headways.

It is possible to demonstrate the use of the Poisson and negative exponential distributions by taking observations of headways on a highway where traffic is flowing freely This flow condition is likely to be found on two-way two-lane highways when the traffic volume in each direction does not exceed 400 veh/h or 800 veh/h on one-way two-lane carriageways and there are no traffic control devices within a distance of approximately 1 km upstream of the point of observation. The time interval between the passage of successive vehicles is noted using a stopwatch and the resulting headways are placed into classes with a class interval of 2 seconds. A smaller class interval is desirable but it is not normally justified unless more accurate means of measuring the headways are available. It will normally be necessary to continue observations for 30 minutes to obtain sufficient headways.

Observations, derived and theoretical values may be tabulated as shown in table 12.1.

TABLE 12.1

Row 1	Headway class
Row 2	Observed frequency of headways in class
Row 3	Observed frequency of headways > lower class limit
Row 4	Row 3 expressed as a percentage
Row 5	Theoretical percentage of headways > lower class limit (equation 12.2 multiplied by 100 where t = 0, 2, 4 . . .)
Row 6	Theoretical frequency of headways > lower class limit (row 5 multiplied by total number of headways observed)
Row 7	Theoretical frequency of headways in class (obtained by difference between successive values of row 6)

It should be noted that the following rows represent observed and theoretical values

row 2 and row 7

row 3 and row 6

row 4 and row 5

An example of observed and theoretical values is given in table 12.2. This table is incomplete as the greatest headway was in the class 64 to 65·9 seconds, but for the sake of brevity only a limited number of values are given.

TABLE 12.2

Row 1	0	2	4	6	8	10	12	14	16
Row 2	34	27	16	13	13	10	6	7	5
Row 3	168	134	107	91	78	65	55	49	42
Row 4	100	80	64	54	46	39	33	29	25
Row 5	100	84	70	58	49	41	34	29	24
Row 6	168	140	117	98	82	68	57	48	40
Row 7	28	23	19	16	14	11	9	8	7

Note that differences between successive values in row 3 give values in row 2; similarly differences in row 6 give values in row 7. This is because the number of headways greater than t_1 second minus the number of headways greater than t_2 second gives the number of headways in class $t_1 t_2$ second.

The closeness of fit of the negative exponential distribution to the observed values can be demonstrated graphically by drawing a histogram of the observed and theoretical number of headways in each class as in figure 12.1.

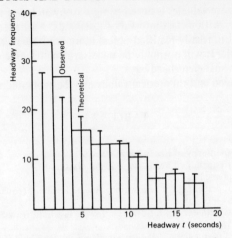

Figure 12.1 A comparison of observed and theoretical headway frequencies

However a statistical test of the closeness of fit of the theoretical distribution can be made by means of the chi-squared test, where chi-squared is the sum for each headway class of the following statistic

$$\frac{(\text{observed frequency} - \text{theoretical frequency})^2}{\text{theoretical frequency}}$$

that is

$$\frac{(\text{row 2} - \text{row 7})^2}{\text{row 7}}$$

Where the theoretical or observed frequency is less than 8 in any class, that class is combined with subsequent classes until the combined theoretical and observed frequencies are greater than 8.

Chi-squared is calculated in table 12.3.

TABLE 12.3   Calculation of closeness of fit

Row 2	Row 7	(Row 2–row 7)^2/row 7
34	28	1·28
27	23	0·69
16	19	0·47
13	16	0·56
13	14	0·07
10	11	0·09
18*	24*	1·50 = 4·66

* Last three classes combined.

Examination of tables of chi-squared shows that at the 5 per cent level the value of chi-squared for 5 degrees of freedom is 11·07. The calculated value is 4·66 showing that there is no significant difference between the observed and theoretical headway distribution at the 5 per cent level.

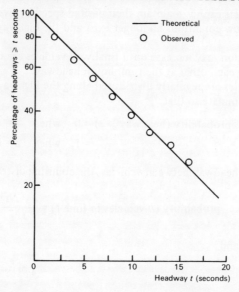

Figure 12.2 A comparison of observed and theoretical cumulative headway distributions

Note that the degrees of freedom are equal to the number of combined classes, in this case 7, minus the number of parameters that were used from the observed to calculate the theoretical values. These two parameters are the total number of observed headways and the mean headway; the former was used in the calculation of row 6 and the latter in the calculation of row 5.

Another way in which the fit of the observed distribution to the theoretical distribution can be demonstrated is by plotting the theoretical and observed cumulative percentage frequency. These values are given in rows 4 and 5 respectively. If equation 12.2 is considered and natural logarithms are taken of both sides of the equation, then

$$\ln [\text{probability (headway} > t)] = -qt \qquad (12.3)$$

When plotted on semi-log paper equation 12.3 is a straight line; the nearer the observed distribution is to a straight line the better is the fit. This is illustrated graphically in figure 12.2.

This experiment shows, for the traffic volume observed and for the class interval chosen, that the distribution of headways may be represented by the exponential distribution. It should be noted that one of the difficulties of the exponential distribution, the surplus of small headways less than 1 second, is hidden by the use of a class interval of 2 seconds.

Congested vehicular headway distributions

While free-flowing vehicular headway distributions may be approximated by the negative exponential distribution in the case of the cumulative headway distribution and by the Poisson distribution in the case of the counting distribution, it is obvious that the distribution of headways will depend on the traffic volume and also on the capacity of the highway. If drivers cannot maintain their desired speed by overtaking slower moving vehicles then free-flow conditions no longer exist and the highway is beginning to show signs of congestion.

Highway congestion may increase until finally all vehicles are travelling at the same speed and following each other at their minimum headway.

In this case vehicles are regularly distributed along the highway and the headway distribution is determinate so that,

$$\text{probability (headway} > t) = 0 \quad \text{when} \quad t \neq \bar{t}$$
$$= 1 \quad \text{when} \quad t = \bar{t}$$

where $\bar{t}$ is the mean headway between vehicles; the counting distribution is also given by

$$\text{probability } (n \text{ vehicles in time } t) = \frac{t - N\bar{t}}{\bar{t}}$$

where $n = N + 1$, and

$$= 1 - \frac{t - N\bar{t}}{\bar{t}}$$

where $n = N$,

N is the number of headways of time headway $\bar{t}$ contained in time interval t.

It is not however so much the two extremes of regularity or randomness which is of interest in the study of traffic flow as the distribution of these headways in the intermediate conditions between these extremes. The study of traffic flow requires a knowledge of the distribution of headways in a wide variety of traffic conditions if realism is to be achieved.

A difficulty of the use of the negative exponential distribution even under free-flow conditions is that the probability of observing a headway increases as the size of the headway decreases. As vehicles have a finite length and a minimum following headway this presents a problem when only a limited number of overtakings are observed. For this reason traffic flow has been described by the use of the displaced exponential distribution. This is a suitable distribution for the description of flow in a single lane of a multilane highway.

Where a small number of low-value headways are observed, such as is the case when a limited amount of overtaking is possible, the Pearson type III or the Erlang distribution may be used to represent the headway frequency distribution.

The Pearson type III distribution was developed by Karl Pearson to deal with a wide variety of statistical data. A random variable such as a headway (t) is said to be distributed as the type III distribution if its probability density function is given by

$$f(t) = \frac{b^a}{\Gamma(a)} t^{a-1} e^{-bt} \qquad 0 < t < \infty \tag{12.4}$$

where $\Gamma(a)$ is known as the gamma function and is defined by

$$\Gamma(a) = \int_0^\infty z^{a-1} e^{-z} \, dz$$

Where headways in a single lane are being observed then any headway less than the minimum following distance cannot be observed and the Pearson type III distribution has to be modified to give

$$f(t) = \frac{b^a}{\Gamma(a)} (t - c)^{a-1} e^{-b(t-c)} \qquad c < t < \infty \qquad (12.5)$$

For equation 12.4 the theoretical distribution may be obtained from

$$\bar{t} = \frac{a}{b} \qquad \text{and} \qquad s^2 = \frac{a}{b^2}$$

For equation 12.5 the theoretical distribution may be obtained from

$$\bar{t} = \frac{a}{b}, \qquad s^2 = \frac{a}{b^2}$$

and c is the minimum observed headway.

The Erlang distribution is a simplified form of the Pearson type III distribution in which the parameter a is an integer. This may be written

$$f(t) = \frac{(qa)^a}{(a-1)!} \times t^{a-1} e^{-aqt} \qquad a = 1, 2, 3 \ldots \qquad (12.6)$$

where q is the rate of flow, the reciprocal of the mean time headway.

The fitting of Pearson III and Erlang distributions to headway data

An example of the fitting of Pearson type III distributions is given below for headway data collected on the Leeds–Bradford highway at Thornbury, a one-way, two-lane highway. Headways were observed using a Marconi Speed Meter connected to a graphical recorder and the grouped data is shown in table 12.4.

The mean and the variance of the headway distribution is first calculated, giving

$$\bar{t} = 3 \cdot 4 \text{ s} \qquad \text{and} \qquad s^2 = 7 \cdot 9 \text{ s}$$

where $\bar{t}$ is the mean time headway and s^2 is the variance of the headways.

As headways are observed in the range 0–0·9 s the frequency curve may be assumed to pass through the origin, a reasonable assumption when it is considered that headways were observed on a one-way two-lane carriageway and frequent overtaking was possible. Using equation 12.5 to calculate the theoretical frequency distribution it is necessary to calculate the value of the parameters a and b.

From the relationship $\bar{t} = a/b$ and $s^2 = a/b^2$ then

$$a = 1 \cdot 499 \qquad \text{and} \qquad b = 0 \cdot 435$$

The value of $\Gamma(a)$ was then calculated using the standard gamma function program available for the Hewlett Packard 9100B desk calculator. Its value was 0·923. The value of the theoretical frequency may then be calculated using equation 12.5 by substituting for t the successive values of the mid-class marks.

TABLE 12.4   Observed and fitted headway distributions

Headway class (seconds)	Observed frequency	Theoretical frequency	
		Equation 12.4 Pearson III	Equation 12.6 Erlang
0–0·9	57	75·3	51·0
1–1·9	99	83·6	85·5
2–2·9	69	69·7	80·0
3–3·9	58	53·3	62·4
4–4·9	44	39·4	45·1
5–5·9	25	27·9	30·8
6–6·9	14	19·7	20·4
7–7·9	13	13·9	13·2
8–8·9	11	9·4	8·4
9–9·9	6	6·6	5·2
10–10·9	3	4·5	3·2
11–11·9	3	2·9	2·0
12–12·9	3	2·1	1·2
13–13·9	0	1·2	0·7
14–14·9	1	0·8	0·4
15–15·9	4	0·4	0·2

Where a desk calculator or tables of the gamma function are not available the Erlang distribution, equation 12.6, may be used as a simpler approximation. The value of the parameter a in this distribution reflects the distribution of headways for a range of traffic flow conditions. When $a = 1$ the distribution becomes negative exponential and when a is infinite complete uniformity of headways results, the value of a thus reflecting flow conditions between free flowing and congested states.

For this example the value of a is chosen as 2 and using equation 6 the theoretical frequency distribution is evaluated and given in table 12.4, together with the observed headway frequencies.

A travelling queue headway distribution model

Miller[4] has described a traffic flow model whereby randomly placed vehicles are moved backwards in time where necessary in order to maintain a constant minimum headway. He derived a bunch length distribution which he verified by observation on a straight three lane section of the A4 between London Airport and Slough. In this case the queue length distribution is given by

$$P(n) = n^{n-1} r^{n-1} \exp{(-rn)}/n!$$

where the parameter r is given by bk, where k is the concentration of vehicles in the traffic stream and b is the mean distance headway of bunched vehicles.

It can be shown that

$$r = Bsq$$

where B is the mean time headway of bunched vehicles,
   s is the ratio of the mean speed of bunched vehicles to the mean speed of all vehicles,
   q is the flow.

Queue length distributions were computed by the author and the theoretical and observed queue lengths compared for flow on the Bradford to Wakefield Road when the traffic flow was 523 veh/h. Good agreement between observed and theoretical distributions was noted.

The double exponential headway distribution model

Schuhl[5] proposed a headway distribution in which vehicles travelling along a highway could be considered to be composed of two types, firstly those who were unable to overtake and were restrained in their driving performance and secondly those drivers who were unrestrained by other vehicles on the highway.

Drivers who are restrained by the action of the driver in front can approach to within a minimum time headway of e and their cumulative headway distribution may be represented by

$$\text{probability (headway} \geq t) = L \exp\left(-(t - e)/(\bar{t}_1 - e)\right) \qquad t \geq e \qquad (12.7)$$
$$= L \qquad\qquad\qquad t \leq e$$

where L is the proportion of restrained vehicles in the traffic stream and $\bar{t}_1$ is the mean headway between restrained vehicles.

Similarly drivers who are not restrained by the vehicle in front have no limitation on the minimum headway, as they are able to overtake, and their cumulative headway distribution may be represented by

$$\text{probability (headways} \geq t) = (1 - L) \exp\left(-t/\bar{t}_2\right) \qquad t \geq 0 \qquad (12.8)$$

where $\bar{t}_2$ is the mean headway between unrestrained vehicles.

On the highway both restrained and unrestrained vehicles are present and so the observed headway distribution is represented by the sum of equations 12.7 and 12.8

$$\text{probability (headways} \geq t) = L \exp\left(-(t - e)/(\bar{t}_1 - e)\right)$$
$$+ (1 - L) \exp\left(-t/\bar{t}_2\right) \qquad t \geq e \qquad (12.9)$$

Some of the earliest research into the fit of the double exponential distribution to observed headways on two-lane urban streets, was carried out by Kell[7]. The theoretical cumulative headway distribution chosen for fitting to observed values was

$$\text{probability (headway} \geq t) = \exp\left(a - t/K_1\right) + \exp\left(c - t/K_2\right)$$

and the relationships of the parameters to the traffic volume were found to be

$$K_1 = 4827 \cdot 9/V^{1 \cdot 024}$$
$$a = -0 \cdot 046 - 0 \cdot 000448 V$$
$$K_2 = 2 \cdot 659 - 0 \cdot 0012 V$$
$$C = \exp\left(-10 \cdot 503 + 2 \cdot 829 \ln V - 0 \cdot 173 (\ln V)^2\right) - 2$$

Similarly Grecco and Sword[7] investigated the headway distribution on a $2\frac{1}{2}$ mile section of US52, a 2 lane 2 way urban bypass around Lafayette, Indiana. They considered that the distribution parameters proposed by Kell were too cumbersome and proposed the following relationships.

Average time headway between restrained vehicles 2·5 s, average time headway between unrestrained vehicles

$$24 - 1·22 \frac{\text{lane volume}}{100}$$

minimum time headway between restrained vehicles 1·0 s, and proportion of restrained vehicles

$$0·115 \times \frac{\text{lane volume}}{100}$$

The author has carried out observations of headway distributions in the West Riding of Yorkshire and has reported[8] the following relationship between traffic volume and the parameters of the double exponential distribution.

For one-way highways with two traffic lanes, the proportion of restrained vehicles was given by 0·00158 volume − 1·04222 when the volume was between 660 and 1295 vehicles/hour.

For two-way highways with a single traffic lane in each direction and with limited opportunities to overtake by entering the opposing traffic lane, the proportion of restrained vehicles was given by 0·00146 volume − 0·52985 when the volume was between 380 and 790 vehicles/hour.

The fitting of a double exponential distribution to headway data

The use of this double exponential distribution may be demonstrated by an analysis of headways on a highway where traffic is experiencing some degree of congestion. Normally this can be expected to occur on a two-lane two-way highway when the traffic volume in each direction exceeds approximately 600 veh/h or on a two-lane one-way

TABLE 12.5   Observed headways

Headway class (seconds)	Observed frequency	Number of headways greater than lower class limits	Percentage of headways greater than lower class limit
0–0·9	57	410	100·0
1–1·9	99	353	86·1
2–2·9	69	254	62·0
3–3·9	58	185	45·3
4–4·9	44	127	31·1
5–5·9	25	83	20·4
6–6·9	14	58	14·4
7–7·9	13	44	11·0
8–8·9	11	31	7·8
9–9·9	6	20	5·1
10–10·9	3	14	3·7
11–11·9	3	11	2·9
12–12·9	3	8	2·2
13–13·9	0	5	1·5
14–14·9	1	5	1·5
15–15·9	4	4	1·2

highway when the traffic volume exceeds approximately 1000 veh/h.

The observed headways given in table 12.4 will be used to illustrate the fitting of the double exponential distribution using a graphical technique. Observed headway frequency, cumulative frequency and percentage cumulative frequency distributions are given in table 12.5.

The distinctive graphical plot on semi-log paper of the cumulative headway distribution when the traffic flow is partly restrained is illustrated in figure 12.3, which should be compared with figure 12.2 (p. 111) where the cumulative headway distribution for free-flowing traffic is plotted.

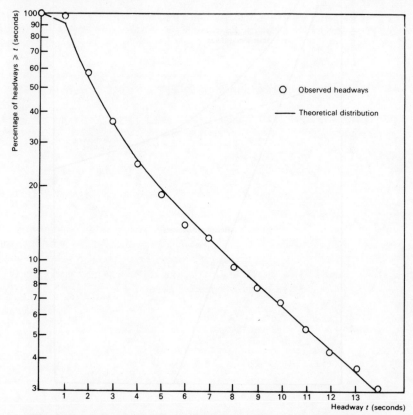

Figure 12.3 Observed headways and the fitted double exponential distribution

The graphical fitting of the double exponential curve may be described by reference to the form of the theoretical curve shown in figure 12.4. The theoretical cumulative headway distribution for both free flowing and restrained vehicles is shown by a solid curve and, if the observed cumulative headway distribution is plotted, it will approximate to this line if the underlying theoretical concept is correct. The straight line portion of this curve represents the headway distribution of free-flowing vehicles as the effect of restrained vehicles is negligible at the larger values of headway. This portion of the curve may then be represented by equation 12.8. If the straight line portion of the curve is extended to the vertical axis, as shown by a broken line, the percentage of headways greater than t represents the value of $100\,(1 - L)$ so allowing L to be determined.

If a point is taken on the line and a value of the probability of a headway $\geqslant t$ and the corresponding value of t are substituted in equation 12.8 then the value of $\bar{t}_2$ may be found.

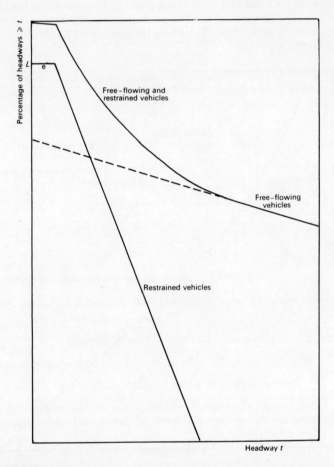

Figure 12.4 The theoretical cumulative headway distribution for free flowing and restrained vehicles in a traffic stream

The vertical difference between the straight line and the cumulative curve represents the headway distribution of restrained drivers. Since this difference represents equation 12.6 it is again a straight line. The value of t, when the proportion of restrained vehicles is L, on this straight line gives e. A point may then be selected on the line and a value of the probability and the corresponding value of t substituted in equation 12.7 to give t.

Using the parameters L, e, $\bar{t}_1$ and $\bar{t}_2$ the theoretical cumulative headway distribution may now be calculated using equation 12.9 and plotted. In this instance the theoretical values were obtained using the least squares technique and the computer program developed by the author RJSE (10) giving the following values of the parameters

$$L \quad 0.66$$

$$t_1 \quad 3.0 \text{ seconds}$$

$$t_2 \quad 4.4 \text{ seconds}$$

It is interesting to note that at the traffic volume sampled, approximately 1000 veh/h, considerably more than 50 per cent of drivers appear to be restrained by the preceding vehicles. Also at a time headway in the region of 7 seconds drivers appear to cease to be influenced by the vehicle in front.

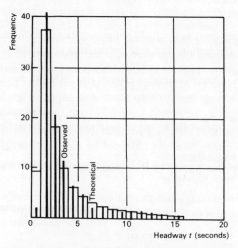

Figure 12.5 Observed and theoretical headway frequencies for congested flow

The observed and the theoretical headway frequencies obtained from the double exponential distribution using the derived parameters are compared in figure 12.5. It can be seen that once again the exponential distribution gives too high a frequency of headways in the class 0–1 second when compared with the observed frequency.

References

1. F. A. Haight, B. F. Whisler and W. W. Mosher. New statistical method for describing highway distribution of cars. *Proc. Highw. Res. Bd*, **40** (1961) 557–64
2. J. P. Kinzer. Applications of the theory of probability to problems of highway traffic. *Proc. Inst. Traffic Engrs* (1934), 118–23
3. W. F. Adams. Road traffic considered as a random series. *J. Instn civ. Engrs* (1936), 121–30
4. A. J. Miller. A queueing model of road traffic flow. *J. R. statist. Soc.*, **23B** (1961), 1.
5. A. Schuhl. The probability theory applied to distribution of vehicles on two-lane highways. *Poisson and Traffic*, the Eno Foundation (1955), p. 59–75
6. J. H. Kell. Analysing vehicular delay at intersections through simulation. *Highw. Res. Bd Bull.* 1 (1962), 28–9

7. W. L. Grecco and E. C. Sword. Predictions of parameters for Schuhl's headway distribution. *Traff. Engng* (Feb. 1968), 36–8

8. R. J. Salter. A simulation of traffic flow at priority intersections. *J. Instn munic. Engrs*, **98** (Oct. 1971), 10, 239–48

Problems

1. Select the correct completion of the following statements.

(a). The time headway distribution between successive highway vehicles may be obtained
(i) from an aerial photograph of a highway,
or (ii) by an observer equipped with a stopwatch standing at the side of the highway.

(b) The counting distribution as applied to highway traffic flow indicates
(i) the distribution of the numbers of vehicles arriving at a point on the highway during fixed periods of time,
or (ii) the distribution of time intervals between the arrival of vehicles,
or (iii) neither

(c) Gap distributions are normally observed by highway traffic researchers because
(i) the counting distributions require a considerable time to collect adequate data,
(ii) the gap distribution may be easily obtained by the use of aerial photographs.

(d) The time headway distribution is of importance to a study of highway traffic flow because
(i) the probability of observing a headway increases as the size of the headway decreases
or (ii) the inverse of the mean time headway is the rate of traffic flow.

TABLE 12.6

Headway class (second)	Observed frequency
0–0·9	19
1–1·9	67
2–2·9	58
3–3·9	29
4–4·9	26
5–5·9	14
6–6·9	17
7–7·9	7
8–8·9	9
9–9·9	6
10–10·9	5
11–11·9	8
12–12·9	4
13–13·9	4
14–14·9	4
15–15·9	3
16–16·9	3
17–17·9	0
18–18·9	2
19–19·9	1

(e) The negative exponential distribution represents
(i) the cumulative headway distribution of freely flowing highway traffic,
or (ii) the cumulative headway distribution of heavily congested highway traffic,
or (iii) the counting distribution of vehicle arrivals on a traffic signal approach.

2. The headway distribution on a two way urban highway is given in table 12.6.
(a) Assume the traffic is free flowing and graphically fit a theoretical distribution to the observed values. Calculate the theoretical frequencies.

(b) Use a histogram to show the closeness of fit of the theoretical and observed distributions and state what conclusions may be drawn from this comparison.

3. The number of vehicles arriving in two minute intervals at a check point on a highway is given in table 12.7.

TABLE 12.7

No. of vehicles arriving in 2 minute intervals	No. of times this number of vehicles was observed
8	4
7	1
6	4
5	9
4	10
3	11
2	12
1	11
0	1

Show that the traffic flow on this highway is random.

4. The traffic flow on a highway is composed of free flowing and following vehicles. The free flowing vehicles have a mean time headway of 8·0 seconds while the following vehicles do not approach closer to the one in front than a time headway of 0·5 second. When the traffic volume was 1000 vehicles per hour it was noted that there were 25 per cent of free flowing and 75 per cent following vehicles.

Calculate:

(a) the proportion of vehicles with a time headway greater than 6 seconds;
(b) the time headway at which only 5 per cent of vehicles are restrained;
(c) the proportion of following vehicles which have a headway greater than 0·25 second.

Solutions

1. (a) The time headway distribution between successive highway vehicles may be obtained by an observer equipped with a stopwatch standing at the side of the highway.

(b) The counting distribution as applied to highway traffic flow indicates the distribution of the numbers of vehicles arriving at a point on the highway during fixed periods of time.

(c) Gap distributions are normally observed by highway traffic researchers because the counting distributions require a considerable time to collect adequate data.

(d) The time headway distribution is of importance to a study of highway traffic flow because the inverse of the mean time headway is the rate of traffic flow.

(e) The negative exponential distribution represents the cumulative headway distribution of freely flowing highway traffic.

2. To graphically fit a theoretical distribution to the observed headway distribution the percentage cumulative headway distribution will be calculated. If the traffic is free flowing then the arrival of vehicles at a point on the highway will be random and the negative exponential distribution will fit the observed cumulative headway distribution. This can be demonstrated by plotting the observed percentage cumulative distribution on semi-log paper. The relationship will be linear and the best straight line can be fitted to the data to obtain the theoretical cumulative distribution from the straight line. The theoretical cumulative distribution can be converted to a cumulative distribution and the differences between successive values will give the theoretical frequencies.

TABLE 12.8

Class (seconds)	Observed frequency	Observed cumulative frequency	Percentage observed cumulative frequency	Percentage theoretical cumulative frequency	Theoretical cumulative frequency	Theoretical frequency
0–0·9	19	286	100·0	100·0	286	56
1–1·9	67	267	93·4	80·3	230	46
2–2·9	58	200	69·9	64·4	184	36
3–3·9	29	142	49·7	51·7	148	29
4–4·9	26	113	39·5	41·5	119	23
5–5·9	14	87	30·4	36·4	104	19
6–6·9	17	73	25·5	26·7	76	15
7–7·9	7	56	19·6	21·4	61	12
8–8·9	9	49	17·1	17·4	50	10
9–9·9	6	40	14·0	14·0	40	8
10–10·9	5	34	11·9	11·2	32	6
11–11·9	8	29	10·1	9·0	26	5
12–12·9	4	21	7·3	7·2	21	4
13–13·9	4	17	5·9	5·8	17	3
14–14·9	4	13	4·5	4·6	13	3
15–15·9	3	9	3·1	3·7	11	2
16–16·9	3	6	2·9	3·0	9	2
17–17·9	0	3	1·0	2·4	7	1
18–18·9	2	3	1·0	1·9	6	1
19–19·9	1	1	–	1·5	4	0

The observed and theoretical frequencies are compared with the help of a histogram. It can be seen that first class is a particularly poor fit as can be expected with the exponential distribution in which the probability of a headway increases as the headway size decreases. This does not occur on the highway where there is a minimum headway between vehicles unless these vehicles are overtaking.

3. The calculation can be carried out in tabular form as shown in table 12.9.

TABLE 12.9

(1) No. of vehicles per 2 minute interval	(2) No. of times observed (f_0)	(3) $(1) \times (2)$	(4) Theoretical probability*	(5) $\Sigma (2) \times 4$ Theoretical frequency (f_t)	(6) x^2 $\dfrac{(f_0 - f_t)^2}{f_t}$
8	4	32	0·015	0·95 ⎫	
7	1	7	0·036	2·27 ⎬	0·18
6	4	24	0·073	4·60 ⎭	
5	9	45	0·128	8·06	0·11
4	10	40	0·187	11·78	0·27
3	11	33	0·218	13·78	0·54
2	12	24	0·191	12·03	0·00
1	11	11	0·110	6·99 ⎫	
0	1	0	0·032	2·02 ⎭	1·04

$\Sigma 63$ $\Sigma 216$ $\Sigma 2·14$

* Probability (n vehicles arriving per 2 minute interval)

$$= \frac{\exp (-m)m^n}{n!}$$

where $m = \Sigma ((1) \times (2))/\Sigma (2) = 3.43$ vehicles.

The number of degrees of freedom is equal to the number of groups (6) minus the constraints (the mean number of vehicles arriving per 2 minute interval, and the total number of vehicle arrivals).

From tables of chi-squared it can be seen that the value of chi-squared for 4 degrees of freedom at the 5 per cent level of significance is 9·49. There is thus no significant difference at the 5 per cent level of significance between the observed and the theoretical distributions indicating that vehicle arrivals are random.

4. The cumulative headway distribution of the freely flowing vehicles in the stream is given by

$$\text{probability } (h > t) = (1 - L) \exp (-t/\bar{t}_2)$$

while the cumulative headway distribution of the following vehicles in the stream is given by

$$\text{probability } (h > t) = L \exp (-(t - e)/(\bar{t}_1 - e))$$

From the information given in the question

$$\text{traffic volume} = \frac{3600}{L\bar{t}_1 + (1 - L)\bar{t}_2}$$

$$\bar{t}_2 = 8·0 \text{ seconds}$$

$$L = 0·75$$

Hence

$$1000 = \frac{3600}{0 \cdot 75t_1 + 0 \cdot 25 \times 8 \cdot 0}$$

that is

$$\overline{t}_1 = 2 \cdot 1 \text{ seconds.}$$

(a) The proportion of vehicles, both free flowing and following, with a time headway greater than 6 seconds is given by

$$\text{probability } (h > 6) = (1 - 0 \cdot 75) \exp (-6/8) + 0 \cdot 75 \exp (-(6 - 0 \cdot 5)/(2 \cdot 1 - 0 \cdot 5))$$
$$= 0 \cdot 25 \times 0 \cdot 4724 + 0 \cdot 75 \times 0 \cdot 1385$$
$$= 0 \cdot 1181 + 0 \cdot 2389$$
$$= 0 \cdot 3570$$

(b) The probability of a headway relating to a following vehicle may be calculated fro

$$\text{probability } (h > t) = L \exp (-(t - e)/(\overline{t}_1 - e))$$

In this case the equation is solved to obtain t when the probability is $0 \cdot 05$

$$0 \cdot 05 = 0 \cdot 75 \exp (-(t - 0 \cdot 5)/(2 \cdot 1 - 0 \cdot 5))$$
$$0 \cdot 067 = \exp (-(t - 0 \cdot 5)/1 \cdot 6)$$

Taking natural logarithms of both sides of the equation

$$-2 \cdot 7031 = - \frac{t - 0 \cdot 5}{2 \cdot 7}$$

$$t = 4 \cdot 8 \text{ seconds.}$$

(c) As all following vehicles have a headway greater than $0 \cdot 5$ second then all following vehicles have a headway greater than $0 \cdot 25$ second. The proportion of following vehicle is $0 \cdot 75$ and so the proportion of following vehicles with a headway greater than $0 \cdot 25$ second is $0 \cdot 75$.

13

The relationship between speed, flow and density of a highway traffic stream

Theoretical relationships between speed, flow and density

When considering the flow of traffic along a highway three descriptors are of considerable significance. They are the speed and the density or concentration, which describe the quality of service experienced by the stream; and the flow or volume, which measures the quantity of the stream and the demand on the highway facility.

The speed is the space mean speed; the density or concentration is the number of vehicles per unit length of highway and the flow is the number of vehicles passing a given point on the highway per unit time.

The relationship between these parameters of the flow may be derived as follows. Consider a short section of highway of length L in which N vehicles pass a point in the section during a time interval T, all the vehicles travelling in the same direction.

The volume flowing $Q = N/T$

The density D $= \dfrac{\text{average no. of vehicles travelling over } L}{L}$

The average number of vehicles travelling over L is given by

$$\frac{\sum\limits_{i=1}^{N} t_i}{T}$$

where t is the time of travel of the ith vehicle over the length L; then

$$D = \frac{\sum\limits_{i=1}^{N} t_i}{T} \bigg/ L = \frac{\dfrac{N}{T}}{L} \qquad \frac{1}{N} \sum\limits_{i=1}^{N} t_i$$

125

or,

$$\text{density} = \frac{\text{volume}}{\text{space mean speed}} \tag{13.1}$$

Numerous observations have been carried out to determine the relationship between any two of these parameters for, with one relationship established, the relationship between the three parameters is determined. Usually the experimenters have been interested in the relationship between speed and volume because of a desire to estimate the optimum speed for maximum flow.

Greenshields[1] is one of the earliest reported researchers in this field and in a study of rural roads in Ohio he found a linear relationship between speed and density of the form

$$\bar{V}_s = \bar{V}_f - \left(\frac{\bar{V}_f}{D_j}\right) D \tag{13.2}$$

where $\bar{V}_s$ is the space mean speed,
$\bar{V}_f$ is the space mean speed for free flow conditions,
D_j is the jam density.

With this relationship determined the volume density relationship can be obtained by substitution of equation (13.1) in equation (13.2) to give

$$Q = \bar{V}_f D - \frac{\bar{V}_f}{D_j} D^2 \tag{13.3}$$

and similarly the relationship between volume and speed may be obtained as

$$Q = D_j \bar{V}_s - \frac{D_j}{\bar{V}_f} \bar{V}_s^2 \tag{13.4}$$

The density and the speed at which volume is a maximum can be obtained by differentiating equations 13.3 and 13.4 with respect to density and speed. To obtain the density when volume is a maximum, from equation 13.3

$$\frac{dQ}{dD} = \bar{V}_f - \left(2 \times \frac{\bar{V}_f}{D_j} D\right) = 0 \text{ for a maximum value}$$

$$D = D_{max} = \frac{D_j}{2} \tag{13.5}$$

To obtain the speed when volume is a maximum, from equation 13.4

$$\frac{dQ}{d\bar{V}_s} = D_j - \left(2 \times \frac{D_j}{\bar{V}_f} \bar{V}_s\right)$$

$$\bar{V}_s = \bar{V}_{max} = \frac{\bar{V}_f}{2} \tag{13.6}$$

Substituting these maximum values in equation 13.1 gives $\bar{Q}_{max}$

$$\bar{Q}_{max} = D_{max} \bar{V}_{max} = D_j \bar{V}_f / 4 \tag{13.7}$$

Observations obtained by Greenshields gave the following values

$$\bar{V}_f = 74 \text{ km/h}$$

$$D_j = 121 \text{ vehicles/km}$$

from equation 13.5 $D_{max} = 61$ vehicles/km
from equation 13.6 $\bar{V}_{max} = 37$ km/h
from equation 13.7 $Q_{max} = 2239$ vehicles/h.

Figures 13.1, 13.2 and 13.3 show the relationships between speed and volume, density and speed and flow and concentration respectively.

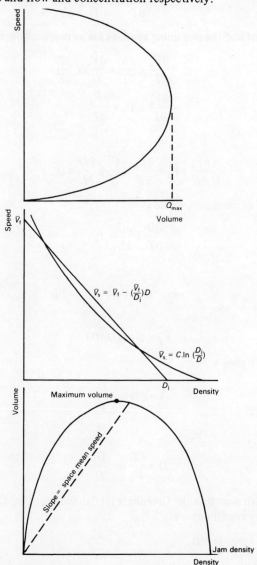

Figure 13.1,2,3 The relationships between speed, volume and density for highway traffic flow

A considerable number of relationships have been proposed between speed and density and the fit of some of these hypotheses to observed data has been investigated by Drake, Schofer and May[2]. Each hypothesis then affects the relationship between speed and volume and also between density and volume.

Greenberg[3] observed traffic flow in the north tube of the Lincoln Tunnel, New York City. He assumed that high density traffic behaved in a similar manner to a continuous fluid for which the equation of motion is

$$\frac{d\bar{V}_s}{dT} = -\frac{C^2}{D} \times \frac{\partial D}{\partial L}$$

where C is a constant and the remaining symbols are as previously defined. Then

$$\frac{d\bar{V}_s}{dT} = \frac{\partial \bar{V}_s}{\partial L} \times \frac{dL}{dT} + \frac{\partial \bar{V}_s}{\partial T} \times \frac{dT}{dT}$$

$$= \frac{\partial \bar{V}_s}{\partial L} \times \bar{V}_s + \frac{\partial \bar{V}_s}{\partial T}$$

or

$$\frac{\partial \bar{V}_s}{\partial L} \times \bar{V}_s + \frac{\partial \bar{V}_s}{\partial T} + \frac{C^2}{D} \times \frac{\partial D}{\partial L} = 0$$

From the equation of continuity of flow

$$\frac{\partial D}{\partial T} + \frac{\partial Q}{\partial L} = 0$$

and

$$Q = \bar{V}_s D$$

then

$$\bar{V}_s = C \ln (D_j/D)$$

Using equation 13.1 gives

$$Q = CD \ln \frac{D_j}{D} \tag{13.8}$$

and

$$D = \frac{CD}{\bar{V}_s} \ln \frac{D_j}{D} \tag{13.9}$$

It is interesting to note that the Greenberg model does not give $\bar{V}_s = \bar{V}_f$. When $D = 0$ the boundary conditions are:

1. When $Q = 0$, $D = 0$ or $D = D_j$.
2. When $D = 0$, $\bar{V}_s = \infty$.
3. When $D = D_j$, $\bar{V}_s = 0$.

The relationship between volume and density illustrated in figure 13.3 has been referred to as the fundamental diagram of traffic[4]. Its form has received considerable attention from those interested in highway flow and it can be described as follows.

The volume flowing is obviously zero when the density is zero and at the jam density the flow may also be assumed to be zero. Between these limits the volume must rise to at least one maximum, often referred to as maximum capacity, to give a shape of the approximate form shown in figure 13.3. At any point on this curve the slope of the line joining that point to the origin is the space speed. The slope is obviously greatest at the origin and decreases to zero at the jam density.

Lighthill and Whitham[5] using a fluid-flow analogy have shown that the speed of waves causing continuous changes of volume through vehicular flow is given by dQ/dD which is the slope of the fundamental diagram, that is

$$V_w = \frac{dQ}{dD}$$

where V_w is the speed of the wave. From equation 13.1

$$V_w = \frac{d(\bar{V}_s D)}{dD} = \bar{V}_s + D \frac{d\bar{V}_s}{dD} \tag{13.10}$$

The space mean speed decreases with increasing density so that $d\bar{V}_s/dD$ is negative. Hence the wave speed is less than that of the traffic stream.

At low densities the second term of equation (13.10) approaches zero and wave velocity approaches the speed of the stream. At the maximum volume the wave is stationary relative to the road since dQ/dD is zero and at higher densities the waves move backwards relative to the stream.

When movement is viewed relative to the road, the wave moves forward at densities less than that of the maximum volume while the wave moves backwards at densities greater than that of the maximum volume.

Changes in density in a traffic stream will result in the production of differing waves travelling at differing velocities through the stream. An example of this could be a length of highway with a heavily trafficked entrance ramp. The wave in the low density length of highway downstream of the entrance ramp will travel forward relative to the highway at a greater speed than the wave in the higher density length of highway upstream of the entrance ramp. When these waves meet, a shock wave will form which will have a velocity

$$V_{sw} = \frac{Q_2 - Q_1}{D_2 - D_1}$$

where Q_1, Q_2 and D_1, D_2 are the respective points on the fundamental diagram for the low and high densities of flow.

A considerable number of observations have been carried out to determine the relationship between speed, volume and density[6,7,8] and the Highway Capacity Manual[9] gives a comprehensive review of the work carried out in the United States.

Observations of the relationship between speed, flow and density

An indication of the relationship between volume, speed and density for a given point on a highway can be obtained by the following observations.

Select a length of highway where the flow is effectively confined to one lane and a range of flow conditions can be expected. By means of a radar speedmeter connected to a graphical recorder or by using a Rustrak or similar recorder take observations of the speeds and time headways of the stream.

The Rustrak recorder is a battery operated device which allows up to four events to be recorded on a moving chart by the depression of one of four push buttons. At least two observers are required, one to mark the entry of a vehicle into a measured speed trap or baseline and the other to mark the exit of the vehicle from the speed trap. If four observers are available more meaningful results can be obtained by taking into account the proportion of commercial vehicles in the flow. In this case two observers would record passenger cars or similar vehicles and two observers would record medium and heavy goods vehicles. Concentration and accuracy are required in this work if confusion between vehicles is to be avoided and in addition some form of time check on the chart is necessary to obtain the relationship between chart distance and real time.

With the time scale of the chart established it is possible to calculate the speed of the vehicles over the length of the baseline by the difference in time between entry and exit. The time headway can similarly be calculated from the difference in time between successive vehicles entering the baseline.

With data obtained from either source the observations should be divided into 5-minute time intervals and the mean values of speed and headway calculated. Note that it is the space mean-speed which is required and not the time mean-speed. The appropriate formula will be

$$\bar{V}_s = \frac{L}{\frac{1}{N}\sum_{i=1}^{N} t_i}$$

where t is the time of travel of the ith vehicle over the measured baseline or speed trap length L when the Rustrak event recorder is used, or

$$\bar{V}_s = \frac{1}{\frac{1}{N}\sum_{i=1}^{N} \frac{1}{V_t}}$$

where V_t is the time mean-speed given by the radar speedmeter.

The mean time headway for each 5 minute interval is obtained by observation and the density is then obtained from

$$D \text{ vehicles/km} = 3600/(\text{mean time headway } \bar{V}_s)$$

where the mean time headway is measured in seconds and $\bar{V}_s$ is the space mean-speed in km/h.

The volume may then be calculated using the relationship given in equation 13.1.

The observations shown in table 13.1 of the traffic flow in the fast lane of the Lincoln Tunnel, New York City by Olcott[11] have been adapted and are used to illustrate the fundamental relationships between density, volume and speed. Observations were made with an event recorder, the speed of vehicles being estimated by time of travel over a measured baseline. Summarised details of the observations are given in table 13.1.

TABLE 13.1

No. of vehicles observed in a 5 min. period	Space mean speed $\bar{V}_s$ (km/h)	Volume Q (veh/h)	Density D (veh/km)	D^2	$\bar{V}_s D$
97	27·0	1164	43·1	1857·6	1163·7
108	25·4	1296	51·0	2601·0	1295·4
104	30·7	1248	40·7	1656·6	1249·5
100	25·6	1200	46·9	2200·0	1200·6
113	34·8	1356	39·0	1521·0	1357·2
116	41·4	1392	33·6	1129·0	1391·0
116	30·2	1392	46·1	2125·2	1392·2
110	40·4	1320	32·7	1069·3	1321·1
115	39·7	1380	34·8	1211·0	1381·6
91	51·2	1092	21·3	453·7	1090·6
	Σ 346·4		Σ 389·2	Σ 15824·4	Σ 12842·9

The fit of this data to the relationship between speed and density proposed by Greenshields and given in equation 13.2 may be shown by estimating $\bar{V}_f$ and $(\bar{V}_f/D_j)$ using the method of least squares. In this method

$$\sum \bar{V}_s = n\bar{V}_f + (\bar{V}_f/D_j) \sum D$$

$$\sum \bar{V}_s D = \bar{V}_f \sum D + (\bar{V}_f/D_j) \sum D^2$$

Substituting the values obtained by summation in table 13.1 gives

$$\bar{V}_f = 71\cdot4 \text{ km/h}$$

$$\frac{\bar{V}_f}{D_j} = -0\cdot94$$

giving

$$D_j = 76 \text{ vehicles/km}$$

The relationship is

$$\bar{V}_s = 71\cdot4 - 0\cdot94D$$

also

$$Q = 71\cdot4D - 0\cdot94D^2$$

and

$$Q = 1\cdot06\bar{V}_s(71\cdot4 - \bar{V}_s)$$

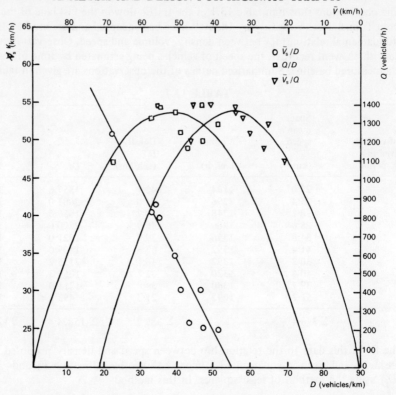

Figure 13.4 Observed and fitted relationships between speed/density, volume/density and speed/volume

These relationships are illustrated graphically in figure 13.4 where observed values and the fitted speed/flow, speed/density and flow/density curves are shown.

From inspection of the graphical plot of flow and space mean-speed, or from differentiation, the maximum flow is found to be 1351 vehicles/h; it occurs when the space mean-speed of the stream is 35·7 km/h.

References

1. B. D. Greenshields. A study of traffic capacity. *Proc. Highw. Res. Bd*, **14** (1934), Pt 1, 448–74

2. J. Drake, J. Schofer and A. May. A statistical analysis of speed–density hypotheses. Vehicular Traffic Science. *Proc. 3rd International Symposium on the Theory of Traffic Flow* (1965), Elsevier, New York (1967)

3. H. Greenberg. An analysis of traffic flow. *Ops Res.* **7** (1959), 1, 79–85

4. F. A. Haight. *Mathematical Theories of Traffic Flow*. Academic Press, New York (1963).

5. M. J. Lighthill and G. B. Whitham. A theory of traffic flow on long roads. *Proc. R. Soc.*, **A229** (1959), 317–45

6. Road Research Laboratory. Research on road traffic H.M.S.O. (1965)
7. J. G. Wardrop. Journey speed and flow in central urban areas. *Traff. Engng Control* **9** (1968), 11, 528-32
8. A. C. Dick. Speed/flow relationships within an urban area. *Traff. Engng Control* Oct. (1966), 393-6
9. Highway Capacity Manual, Highway Research Board Special Report 87, 1965
10. R. J. Salter. Vehicle speeds and headways at northern end of Ml. *Highw. Traff. Engng* (Sept. 1969), 34-6
11. E. A. Olcott. The influence of vehicular speed and spacing on tunnel capacity. Presented at the Informal Seminar in Operations Research, Johns Hopkins University, November 1954

Problems

1. Greenshields proposed a linear relationship between speed and density. Using this relationship it was noted that on a length of highway the free speed $\bar{V}_f$ was 80 km/h and the jam density $\bar{D}_j$ was 70 vehicles/km.

(a) What is the maximum flow which could be expected on this highway?
(b) At what speed would it occur?

2. Observations of the speed and flow through a highway tunnel showed that the relationship between speed and density was of the form

$$\bar{V}_s = 35 \cdot 9 \ln \frac{180}{D}$$

where speeds are measured in km/h and densities in vehicles/km.

(a) What is the jam density on this highway?

3. The diagram below (Figure 13.5) shows the relationship between volume and density for traffic flow on a highway. Four points on the curve are marked A, B, C and D; indicate for the traffic flow situations given below those which could be represented by point(s) A and/or B and/or C and/or D.

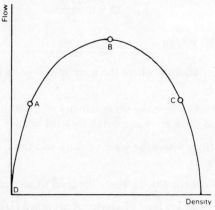

Figure 13.5

(a) Traffic flow conditions where the wave velocity was moving backwards relative to the roadway.

(b) Traffic flow conditions where the wave velocity was less than the space mean-speed of the traffic stream and the wave was moving forward relative to the roadway.

(c) Traffic flow conditions where the wave velocity and the space mean-speed stream velocity were equal.

(d) Congested traffic flow conditions where level of service E prevailed.

(e) Free flow conditions where level of service A prevailed.

Solutions

1. (a) The maximum flow $\bar{Q}_{max}$ is given by

$$D_j \bar{V}_f / 4$$

when there is a linear relationship between speed and density, that is

$$\bar{Q}_{max} = \frac{70 \times 80}{4} = 1400 \text{ veh/h}$$

(b) The speed $\bar{V}_{max}$ of which the flow is a maximum is given by

$$\frac{\bar{V}_f}{2} = 40 \text{ km/h}$$

2. The jam density D_j on a highway is given by

$$\bar{V}_s = C \ln \frac{D_j}{D}$$

Hence in the equation given

$$\bar{V}_s = 35 \cdot 9 \ln \frac{180}{D}$$

the jam density D_j is 180 veh/km.

3. (a) Traffic flow conditions where the wave velocity was moving backwards relative to the roadway (C).

(b) Traffic flow conditions where the wave velocity was less than the space mean-speed of the traffic stream and the wave was moving forward relative to the roadway (A).

(c) Traffic flow conditions where the wave velocity and the space mean-speed stream velocity were equal (D).

(d) Congested traffic flow conditions where the level of service E prevailed (C).

(e) Free flow conditions where level of service A prevailed (D).

14

The distribution of vehicular speeds in a highway traffic stream

Space mean and time mean-speed

One of the fundamental parameters for describing traffic flow is the speed, either of individual vehicles or of the traffic stream. It is of importance in work connected with the theory of traffic flow because of the fundamental connection between speed, flow and concentration when the movement of a traffic stream is being considered.

In traffic management a knowledge of the speed of vehicles is required for the realistic design of traffic signs, the layout of double white lines and the assessment of realistic speed limits. Geometric design of highways also requires that deceleration and acceleration and sight distances, superelevation and curvature must be related to an assumed design speed.

Because speed measurements are used for many purposes in highway traffic engineering and highway design, several differing definitions of speed are commonly used. Where vehicle speed is used to assess journey times, in connection with transportation studies that are concerned with travel over an area, it is the average journey speed that is important. On the other hand if vehicle speed is being used to investigate the accident potential of a section of highway then it is the spot speed which is required. The difference between these two types of speed can be illustrated by a simple example.

Five vehicles travel at 40, 50, 60, 70 and 80 km/h over a distance of 1 km. The mean of the spot speeds is 60 km/h. The mean travel time of the five vehicles is however 0·018 h so that the mean journey speed is 55·6 km/h. When speeds are measured at one point in space over a period of time it is the time mean-speed that is obtained. These individual speeds are often referred to as spot speeds. It is defined as

$$\bar{V}_t = \frac{\sum V_t}{n}$$

where $\bar{V}_t$ is the time mean speed,
   V_t are the individual speeds in time,
   n is the number of observations.
   Where speeds are averaged over space as is the case when the mean journey speed is calculated then it is the space mean-speed that is calculated, that is

$$\bar{V}_s = \frac{n}{\sum 1/\bar{V}_t}$$

where $\bar{V}_s$ is the space mean-speed,
   n is the number of observations.

Wardrop[1] has shown that time and space-mean speeds are connected by the relationship

$$\bar{V}_t = \bar{V}_s + \frac{\sigma_s^2}{\bar{V}_s}$$

where σ_s is the standard deviation of $\bar{V}_s$.

Speed measurements

Most of the speed measurements made by highway traffic engineers produce a time mean-speed because they are obtained by the use of radar speedmeters or timing devices using short baselines. If in the latter case however the mean travel time is used to calculate the mean speed then it is the space mean-speed that is obtained. Speed measurements derived from successive aerial photographs can however be used to derive space mean-speed distributions directly by noting the travel distance of vehicles between successive exposures.
   The simplest and cheapest method of obtaining speed data is to measure the time of travel over a measured distance or baseline. The accuracy of the method depends on the method of timing and where a stopwatch is used it is often assumed that a skilled observer can read to 0·2 s but difficulties of a constant reaction time often mean that the level of accuracy is approximately 0·5 s. The Traffic Engineering Handbook[2] suggests baseline lengths of

   27 m where stream speed is below 40 km/h
   54 m where stream speed is between 40 and 70 km/h
   81 m where stream speed is greater than 70 km/h.

Some of the observer errors associated with timing over a measured length may be overcome by the use of the enoscope, which bends the line of sight of the observer by means of a mirror and eliminates parallax error.
   Where details of journey times are required then the time of travel over distances of 1 to 2 km may be obtained by the use of observers stationed at each end of this extended baseline and having synchronised stopwatches. The time of arrival of each

vehicle into the measured length and the time of its departure are recorded together with some portion of the registration number. Subsequent comparison of times and numbers allows the journey times to be calculated. It is difficult for one observer to read and record more than one registration number every 5 s, even when only a portion of the registration number is recorded. For this reason only registration numbers ending with given digits are recorded by both observers according to the sample size required.

A more realistic form of device for use in present day traffic conditions makes use of two pneumatic tubes attached to the road surface over a short baseline. Most of these instruments are transistorised allowing operation from portable power sources and some allow sampling of vehicle speeds to be carried out, a necessity in heavy traffic flow conditions. As in any observations sampling of the traffic flow should be carried out with care to avoid obtaining a biased sample. Care should be exercised to avoid selecting the first vehicle in a platoon and conversely selection of free flowing single vehicles will tend to bias the resulting speed distribution towards the faster vehicles. Where speed observations are not classified into vehicle type then the combined observations should be in proportion to the traffic composition.

A speed measuring device that does not rely on timing the passage of a vehicle over a base line is the radar speedmeter. These devices operate on the fundamental principle that a radio wave reflected from a moving target has its frequency changed in proportion to the speed of the moving object. Radar meters have been extensively employed both for police and for research purposes. In these meters the radar beam may be transmitted directly along the highway making it necessary to site the meter on a curve or on the central reservation or, as is more usual in the newer models, a beam is transmitted at an angle across the carriageway.

A frequent choice of police forces and traffic engineering departments is the Marconi Portable Electronic Traffic Analyser. This apparatus has an aerial array in which the energy from the transmitter is beamed at an angle of 20° to the main axis of the meter. This means that if the instrument is aimed straight down the road then the radio waves cross the paths of vehicles making it possible to distinguish between individual vehicles.

For police purposes the passage of a vehicle through the beam occurs so quickly that accurate observation of the meter may be difficult. To overcome this difficulty a hold device may be used which causes the meter to remain at the indicated value for a predetermined short time. When used for traffic engineering purposes the speed of successive vehicles may be estimated without the use of this hold device. A graphical recorder may be coupled to the radar meter to obtain a continuous record of vehicle speeds over a considerable period of time. Accuracy of the meter is ±3·2 km/h at speeds up to 130 km/h. Transmitting frequency is in the band 10 675 to 10 699 MHz with a power consumption of 3·5 A at 12 V. A difficulty encountered in the use of this meter with an automatic recording device is the inability to differentiate between vehicles travelling in opposite directions.

Analysis of speed studies

Because in any speed study a considerable number of speeds are observed, statistical techniques are used to analyse the data obtained. Depending upon the accuracy of the data, the use to which the derived results are to be put and the number of observations obtained, a suitable class interval is chosen.

TABLE 14.1

1	2	3	4	5	6	7	8
Speed class (km/h)	Frequency	Percentage frequency	Cumulative frequency	Percentage cumulative frequency	Deviation	$(2) \times (6)$	$(2) \times (6)^2$
44–47·9	1	0·286	1	0·286	−9	−9	81
48–51·9	2	0·571	3	0·857	−8	−16	128
52–55·9	2	0·571	5	1·429	−7	−14	98
56–59·9	4	1·143	9	2·571	−6	−24	144
60–63·9	11	3·143	20	5·714	−5	−55	275
64–67·9	24	6·857	44	12·571	−4	−96	384
68–71·9	40	11·429	84	24·000	−3	−120	360
72–75·9	48	13·714	132	37·714	−2	−96	192
76–79·9	63	18·000	195	55·714	−1	−63	63
80–83·9	40	11·429	235	67·143	0	0	0
84–87·9	34	9·714	269	76·857	1	34	34
88–91·9	29	8·286	298	85·143	2	58	116
92–95·9	25	7·143	323	92·286	3	75	225
96–99·9	13	3·714	336	96·000	4	52	208
100–103·9	5	1·429	341	97·429	5	25	125
104–107·9	3	0·857	344	98·286	6	18	108
108–111·9	1	0·286	345	98·571	7	7	49
112–115·9	2	0·571	347	99·143	8	16	128
116–119·9	2	0·571	349	99·714	9	18	162
120–123·9	1	0·286	350	100·000	10	10	100
	$\Sigma\,350$					$\Sigma\,-180$	$\Sigma\,2980$

Table 14.1 shows speed observations obtained on a major traffic route. Individual speeds have been grouped into 4 km/h classes given in column 1—an interval which reduces the data into an easily managed number of classes yet does not hide the basic form of the speed distribution. In the selection of class intervals thought should be given to the dial readings when visual observation of the speed is made. Most speeds will be recorded to the nearest dial reading and these form convenient mid-class marks.

The number of observations in each class, or the frequency, is given in column 2 and converted into the percentage in each class by dividing the individual values in column 2 by the sum of column 2. This percentage frequency is given in column 3. The cumulative number of observations or cumulative frequency is given in column 4. This column represents the number of vehicles travelling at a speed greater than the lower class limit. In column 5 the percentage cumulative frequency is given. It is obtained by dividing the value in column 4 by the total number of speeds observed.

It is often assumed that speeds are normally distributed. To test this hypothesis it is necessary to estimate the mean speed and the standard deviation of the observed speeds. Both the mean and standard deviation can be calculated by the use of coding to reduce the arithmetic manipulation necessary. A class is selected which is considered likely to contain the mean speed, although it is not essential that the mean does lie within the class. The number of class deviations from this selected class is given in column 6. The sign of the deviation should be particularly noted.

These deviations are multiplied by the corresponding frequency, given in column 2, and the resulting value is entered in column 7.

The mean speed is then given by

$$\text{mid-class mark of selected class} + \frac{\text{class interval } \Sigma \, (\text{column } 7)}{\Sigma \, (\text{column } 2)}$$

$$82 - \frac{4 \cdot 180}{350} = 79 \cdot 9 \text{ km/h}$$

The standard deviation is given by

$$\text{class interval} \sqrt{\left[\frac{\Sigma \, (\text{frequency } (\text{deviation})^2)}{\Sigma \, (\text{column } 2)} - \left(\frac{\Sigma \, (\text{frequency} \times \text{deviation})}{\Sigma \, (\text{column } 2)}\right)^2\right]}$$

The value of Σ (frequency × deviation) has already been calculated in column 6 and it is now necessary to calculate the frequency (deviation)^2 for each speed class. These values are given in column 8.

$$4 \sqrt{\left[\frac{2980}{350} - \left(\frac{-180}{350}\right)^2\right]} = 11 \cdot 6 \text{ km/h}$$

The two parameters of the normal distribution have now been determined and it is possible to calculate the theoretical values of speed frequency in each class by either of two methods. These are:

1. by the use of a table of areas under the normal probability curve.
2. by the use of the probit method as developed by Finney[3].

Using tables of the area under a normal probability curve it is possible to calculate the theoretical frequency assuming a normal distribution.

This is performed in a tabular manner in table 14.2. In column 1 the upper class limits used in the original speed observations of table 14.1 are given. The deviation of these class limits from the previously calculated mean are given in column 2 and then converted into standard deviations from the mean by dividing the value in column 2 by the previously calculated standard deviation. These values are given in column 3. From tables of the normal probability curve. the area under a normal curve between these class limits and the mean can be obtained and these values are given in column 4. The difference between successive values in column 4 gives the theoretical area under the normal curve between the class limits. This is the theoretical probability of a speed lying between the class limits and is given in column 5. When this probability is multiplied by the observed total frequency given in table 14.1 the theoretical frequency is obtained. This is given in column 6.

The second method, which uses the probit analysis approach, is based on the percentage cumulative frequency, given in column 5, table 14.1. This approach is described by Finney[2] and uses the relationship that when speeds are distributed normally the cumulative speed distribution may be written

Percentage of vehicles travelling at a speed equal to, or less than,

$$V = 1/(\sigma \times 2\pi) \int_{-\infty}^{V} \exp\left(-(V - \bar{V})^2/2\sigma^2\right) \, \mathrm{d}v$$

where $\bar{V}$ is the mean speed,
$\quad \sigma$ is the standard deviation of the speeds.

A demonstration of the fit of the observed cumulative speed distribution to a cumulative normal distribution may be obtained by plotting the probit of the percentage of

TABLE 14.2

1	2	3	4	5	6	7	8
Upper speed class limit (km/h)	Column 1 minus mean speed	Column 2 divided by standard deviation	Normal area	Probability	Theoretical frequency	Observed frequency	$\dfrac{((6)-(7))^2}{(6)}$
44	−35·9	−3·10	−0·499				
48	−31·9	−2·75	−0·497	0·002	0·7 ⎫	1 ⎫	
52	−27·9	−2·40	−0·492	0·005	1·8 ⎪	2 ⎪	
56	−23·9	−2·06	−0·480	0·012	4·2 ⎬	2 ⎬	2·27
60	−19·9	−1·72	−0·457	0·023	8·1 ⎭	4 ⎭	
64	−15·9	−1·37	−0·415	0·042	14·7	11	0·93
68	−11·9	−1·025	−0·349	0·066	23·1	24	0·04
72	−7·9	−0·680	−0·252	0·097	33·9	40	1·10
76	−3·9	−0·336	−0·132	0·119	41·9	48	0·89
80	+0·1	0·009	0·004	0·137	48·0	63	4·69
84	+4·1	0·354	0·138	0·134	46·9	40	1·02
88	+8·1	0·70	0·258	0·120	42·0	34	1·52
92	+12·1	1·04	0·351	0·093	32·6	29	0·40
96	+16·1	1·39	0·418	0·067	23·4	25	0·11
100	+20·1	1·74	0·459	0·041	14·3 ⎫	13 ⎫	0·12
104	+24·1	2·08	0·481	0·022	7·7 ⎪	5 ⎪	
108	+28·1	2·42	0·492	0·011	3·8 ⎪	3 ⎪	
112	+32·1	2·76	0·497	0·005	1·8 ⎬	1 ⎬	0·01
116	+36·1	3·11	0·499	0·002	0·7 ⎪	2 ⎪	
120	+40·1	3·46	0·500	0·001	0·4 ⎪	2 ⎪	
124	+44·1	3·81	0·500	0·000	0 ⎭	1 ⎭	

$$\Sigma\ 13\cdot10$$

vehicles travelling at or less than a certain speed, against the speed upper class limit. Values of probits may be obtained from the suggested reading or can be obtained from Figure 14.1. The use of this technique converts a cumulative normal curve into a straight line whose equation is

Probit of percentage of vehicles travelling at a speed $< V = 5 + \dfrac{1}{\sigma}\,(V - \bar{V})$

Using the derived values of σ and $\bar{V}$ this gives

Probit of percentage of vehicles travelling at a speed $< V = 5 + 0 \cdot 0862\,(V - 79.9)$

$$= 0 \cdot 0862\,V - 1 \cdot 6887$$

$$(14.1)$$

This line is plotted in figure 14.1 together with the observed cumulative frequency distribution obtained from column 5 of table 14.1. The close agreement except at high speeds can be seen.

Using the equation above it is possible to calculate the theoretical probit of vehicles travelling at a speed $> V$, and using either the scale given in figure 14.1 or the tables

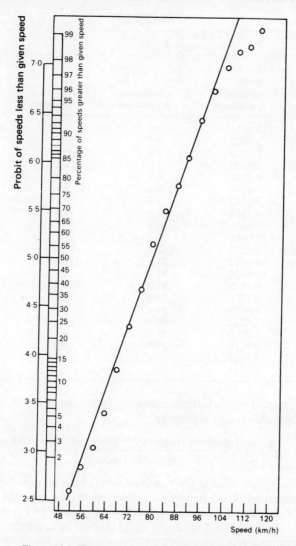

Figure 14.1 The observed and fitted speed distribution

given by Finney it is possible to calculate the theoretical percentage acceptance. This is done in columns 2 and 3 of table 14.3. The difference between successive values in column 3 gives the percentage frequency, which is given in column 4. When this column is multiplied by the total number of speeds observed the frequency is obtained and this is given in column 5. The values given in this column differ slightly from those derived in column 6 of table 14.2. This is due to errors involved in the interpolation of values in the tables used in the calculation.

It is often necessary to calculate whether the observed speed frequencies differ significantly from the expected speed frequencies. It is possible to estimate whether the

TABLE 14.3

1	2	3	4	5
Speed class limit (km/h)	Probit cumulative percentage	Cumulative percentage	Percentage frequency	Frequency
44	2·1031	0·0		
48	2·2502	0·3	0·3	0·1
52	2·5950	0·8	0·5	1·8
56	2·9398	2·0	1·2	4·2
60	3·2746	4·3	2·3	8·1
64	3·6294	8·6	4·3	15·1
68	3·9742	15·3	6·7	23·5
72	4·3190	24·8	9·5	33·3
76	4·6638	36·9	12·1	42·4
80	5·0086	50·3	13·4	46·9
84	5·3534	63·8	13·5	47·3
88	5·6982	75·8	12·0	42·0
92	6·0430	85·2	9·4	32·9
96	6·3878	91·7	6·5	22·8
100	6·7326	95·9	4·2	14·7
104	7·0774	98·1	2·2	7·7
108	7·4222	99·2	1·1	3·9
112	7·7670	99·7	0·5	1·8
116	8·1118	99·9	0·2	0·7
120	8·4566		0·1	0·4
124	8·8014			

observed and expected results differ significantly by the use of the chi-squared test of significance where chi-squared is given by

$$\chi^2 = \sum \frac{(\text{observed frequency} - \text{expected frequency})^2}{\text{expected frequency}}$$

If χ^2 exceeds the critical value at the 5 per cent level of significance then it is said that there is a significant difference between the observed and the expected values.

The value of chi-squared is calculated in column 8 of table 14.2 and summed at the bottom of the column to give a value of 13·10. It should be noted that where an individual frequency is less than 8 both the observed and the expected frequencies have been combined with other classes.

The critical value of chi-squared which must not be exceeded depends on the number of degrees of freedom. This is given by the number of rows in column 8 (since the greater the number of individual values forming the sum of chi-squared the greater will be its expected value) minus the constraints or the number of parameters which were taken from the observed frequency values to calculate the theoretical frequency values. In this case the constraints are the mean speed, the standard deviation of speeds and the total frequency of speeds. This means the number of degrees of freedom of the calculated value of chi-squared is $12 - 3 = 9$.

Tables of chi-squared given in most statistical text books show that the critical value of chi-squared for 9 degrees of freedom at the 5 per cent level of probability is 15·5.

The calculated value is 13·10 and so it can be assumed that the observed speed distribution may be represented by a normal distribution.

Having established that speeds are normally distributed it is possible to use the well-known properties of the normal distribution to obtain detailed information on the speed distribution. For example, it is often desirable to calculate the 85 percentile speed and this can be obtained by the use of equation 14.1. The probit of 85 per cent is 6·0364 giving

$$6·0364 = 0·0862V - 1·6887$$

or

$$V(85 \text{ percentile}) = 89·6 \text{ km/h}$$

It is also a property of the normal distribution that 68·27 per cent of all observations will lie within plus or minus one standard deviation of the mean value. This means that approximately two-thirds of all vehicles will be travelling at a speed between 79·9−11·6 km/h and 79·9 + 11·6 km/h, that is between 68·3 and 91·5 km/h. Examinations of column 2 of table 14.1 shows that this is correct.

A knowledge of the mathematical form of the speed distribution is also of considerable importance in many theoretical studies of traffic flow and in the simulation of driver behaviour in many highway situations. It has also been suggested[4] that an examination of the speed distribution at a particular location will give an indication of the accident potential of the highway. Where the distribution is normal it is suggested that the accident potential is less than when there is an undue proportion of high or low speeds.

References

1. J. G. Wardrop. Some theoretical aspects of road traffic research. *Proc. Instn civ. Engrs*, 2 (Pt 1) (1952), 2, 325–62
2. Institute of Traffic Engineers, Traffic Engineering Handbook (1965)
3. D. J. Finney. Probit analysis, a statistical treatment of the sigmoid response curve, Cambridge University Press (1947)
4. W. C. Taylor. Speed zoning: a theory and its proof. *Traff. Engng.* 35 (1965), 4, pp 17–19, 48, 50–1

Problems

1. Drivers in a vehicle testing programme travel around a race track at a constant speed measured by radar speedmeters and by timing their travel time around the track. The mean speed of all the vehicles was obtained by the two methods and compared.

(a) If all the vehicles travel at the same speed as indicated by their corrected speedometers, will the

(i)   two mean speeds be the same?
(ii) the radar speedmeter mean be higher?
(iii) the radar speedmeter mean be the lower?

(b) If the vehicles travel at differing speeds as indicated by their corrected speedometers will the

(i)   two mean speeds be the same?
(ii) the radar speedmeter mean be higher?
(iii) the radar speedmeter mean be the lower?

2. The speed distribution on a rural trunk road was noted to be normally distributed with a mean speed of 90 km/h and a standard deviation of 20 km/h. Using the probit technique estimate the 85 percentile speed on the highway.

Solutions

1. When the speeds are measured by a radar speedmeter and averaged it is the time mean-speed $\bar{V}_t$ that is obtained whereas when speeds are measured from travel time it is the space mean-speed $\bar{V}_s$ that is calculated. The relationship between the two mean speeds is given by

$$\bar{V}_t = \bar{V}_s + \frac{{\sigma_s}^2}{\bar{V}_s}$$

(a) In this case all the vehicles travel at the same speed and the standard deviation of the space mean-speeds σ_s is zero so that the two mean speeds are the same.

(b) In this case the vehicles travel at differing speeds and since the standard deviation of speeds is positive the mean speed obtained from the radar speedmeter will be higher than the mean speed obtained from journey times.

2. When speeds can be represented by the cumulative normal distribution then the following relationship may be used.

$$\text{Probit of percentage of vehicles travelling at a speed} > V = 5 + \frac{1}{\sigma}(V - \bar{V})$$

Using the values given

$$\text{Probit of percentage of vehicles travelling at a speed} > V = 5 + \frac{1}{20}(V - 90)$$

The probit of 85 per cent is 6·03, that is

$$6 \cdot 03 = 5 + \frac{1}{20}(V_{85} - 90)$$

or

$$V_{85} = 110 \cdot 6 \text{ km/h}$$

15

The macroscopic determination of speed and flow of a highway traffic stream

The relationships considered in chapter 13 between speed, flow and density of a traffic stream considered traffic flow in a microscopic sense in that the speeds of, and headways between, individual vehicles have been used to obtain relationships for the whole traffic stream. An alternative approach to the determination of speed/flow relationships is the use of macroscopic relationships obtained by the observation of stream behaviour.

Speed and flow of a moving stream of vehicles may be obtained by what has become known as the moving car observer method. Wardrop and Charlesworth[1] have described how stream speed and flow may be estimated by travelling in a vehicle against and with the flow. The journey time of the moving observation vehicle is noted, as is the flow of the stream relative to the moving observer. This means that when travelling against the flow the relative flow is given by the number of vehicles met and while travelling with the flow the relative flow is given by the number of vehicles that overtakes the observer minus the number that the observer overtakes. Then

$$Q = \frac{(x + y)}{(t_a + t_w)}$$

and

$$\bar{t} = t_w - y/Q$$

where Q is the flow,
  $\bar{t}$ is the mean stream-journey time,
  x is the number of vehicles met by the observer, when travelling against the stream,
  y is the number of vehicles that overtakes the observer minus number overtaken,
  t_a is the journey time against the flow,
  t_w is the journey time with the flow.

145

These relationships are determined as follows. Consider a stream of vehicles moving along a section of road, of length l, so that the average number Q passing through the section per unit time is constant. This stream may be regarded as consisting of flows

$$Q_1 \text{ moving with speed } V_1$$

$$Q_2 \text{ moving with speed } V_2$$

$$\text{etc.} \qquad \text{etc.}$$

Suppose an observer travels with the stream at speed V_w and against the stream at speed V_a, then the flows relative to him, of vehicles with flow Q_1 and speed V_1 are

$$Q_1(V_1 - V_w)/V_1$$

and

$$Q_1(V_1 + V_a)/V_1$$

respectively.

If t_w is the journey time $1/V_w$, t_a is the journey time $1/V_a$ and t_1 is the journey time $1/V_1$, etc., then

$$Q_1(t_a + t_1) = x_1 \qquad \text{and} \qquad Q_2(t_a + t_2) = x_2, \text{etc.}$$

$$Q_1(t_w - t_1) = y_1 \qquad \text{and} \qquad Q_2(t_w - t_2) = y_2, \text{etc.}$$

where x_1, x_2, etc., are the number of vehicles travelling at speeds V_1, V_2, etc. met by the observer when travelling against the stream and y_1, y_2, etc., are the number of vehicles overtaken by the observer minus the number overtaking the observer.

Summing over x and y

$$x = x_1 + x_2 + x_3 + \cdots$$

$$y = y_1 + y_2 + y_3 + \cdots$$

$$x = Q_1(t_a + t_1) + Q_2(t_a + t_2) + \cdots$$

$$y = Q_1(t_w - t_1) + Q_2(t_w - t_2) + \cdots$$

$$x = Qt_a + \Sigma(Q_1t_1 + Q_2t_2 + \cdots)$$

$$y = Qt_w - \Sigma(Q_1t_1 + Q_2t_2 + \cdots)$$

and

$$\bar{t} = \frac{(Q_1t_1 + Q_2t_2 + Q_3t_3 + \cdots)}{Q}$$

so that

$$x = Qt_a + Q\overline{t}$$

$$y = Qt_w - Q\overline{t}$$

or

$$Q = (x + y)/(t_a + t_w)$$

and

$$\overline{t} = t_w - y/Q$$

so allowing the stream speed to be calculated.

The application of this method of measuring speed and flow is illustrated by the following field data, which was obtained to estimate the two-way flow on a highway. To reduce the number of runs required by the observation vehicles, details of the relative flow of both streams are obtained for each run of the observer's vehicle. For this reason, details are given of two relative flows in the data obtained when the observer is travelling eastwards and when he is travelling westwards. These observations are given in table 15.1.

Observer travelling to east   **TABLE 15.1**

Line	Time of commencement of journey	Journey time	No. of vehicles met	No. of vehicles overtaking observer	No. of vehicles overtaken by observer
1	09·20	2·51	42	1	0
2	09·30	2·58	45	2	0
3	09·40	2·36	47	2	1
4	09·50	3·00	51	2	1
5	10·00	2·42	53	0	0
6	10·10	2·50	53	0	1

Observer travelling to west

Line	Time of commencement of journey	Journey time	No. of vehicles met	No. of vehicles overtaking observer	No. of vehicles overtaken by observer
7	09·25	2·49	34	2	0
8	09·35	2·36	38	2	1
9	09·45	2·73	41	0	0
10	09·55	2·41	31	1	0
11	10·06	2·80	35	0	1
12	10·15	2·48	38	0	1

Length of highway test section 1.6 km.

The information relating to the *highway flow to the east* is abstracted and tabulated in table 15.2.

TABLE 15.2

Time of commencement of journey	Relative flow rate		t_a	t_w	Q (veh/min)	$\bar{t}$ (min)	V (km/h)
	with observer	against observer					
09·20	1			2·51			
09·25		34	2·49		7·0	2·37	41
09·30	2			2·58			
09·35		38	2·36		8·1	2·33	42
09·40	1			2·36			
09·45		41	2·73		8·3	2·24	43
09·50	1			3·00			
09·55		31	2·41		6·3	2·84	34
10·00	0			2·42			
10·05		35	2·89		6·6	2·42	40
10·10	−1			2·50			
10·15		38	2·48		7·4	2·36	41

Similarly, the information relating to the highway flow to the west is abstracted and tabulated in table 15.3.

TABLE 15.3

Time of commencement of journeys	Relative flow rate		t_a (min)	t_w (min)	Q (veh/min)	$\bar{t}$ (min)	V (km/h)
	with observer	against observer					
09·20		42	2·51				
09·25	2			2·49	8·8	2·38	41
09·30		45	2·58				
09·35	1			2·36	9·3	2·25	43
09·40		47	2·36				
09·45	0			2·73	9·3	2·62	37
09·50		51	3·00				
09·55	1			2·41	9·6	2·31	42
10·00		53	2·42				
10·05	−1			2·89	9·8	2·79	35
10·10		53	2·50				
10·15	−1			2·48	10·4	2·39	40

If the mean stream speed and flow is required then these individual values may be averaged, to give

stream velocity = 40 km/h

stream flow = 9·5 veh/min

Reference

1. J. G. Wardrop and G. Charlesworth. A method of estimating speed and flow of traffic from a moving vehicle. *J. Instn civ. Engrs*, **3** (Pt 2) (1954), 158–71

Problem

In a stream of vehicles 30 per cent of the vehicles travel at a constant speed of 60 km/h, 30 per cent at a constant speed of 80 km/h and the remaining vehicles travel at a constant speed of 100 km/h. An observer travelling at a constant speed of 70 km/h with the stream over a length of 5 km is passed by 17 vehicles more than he passes. When the observer travels against the stream at the same speed and over the same length of highway the number of vehicles met is 303.

(a) What is the mean speed and flow of the traffic stream?
(b) Is the time mean or the space mean-speed obtained by this technique?
(c) How many vehicles travelling at 100 km/h pass the observer, while he travels with the stream?

Solutions

1. (a) The flow Q of a traffic stream may be obtained from

$$\frac{x+y}{t_a + t_w}$$

The time of travel t_a and t_w is 5 km per 70 km/h. Then

$$Q = \frac{303 + 17}{5/70 + 5/70}$$

$$= 2240 \text{ vehicles/h}$$

and

$$\bar{t} = \frac{5}{70} - \frac{17}{2240}$$

$$= 0 \cdot 0714 - 0 \cdot 0076$$

$$= 0 \cdot 0638$$

The mean speed of the stream = $5/0 \cdot 0638$ km/h

$$= 78 \cdot 3 \text{ km/h}$$

(b) The speed obtained by the moving observer technique is calculated from a journey time and hence it is the space mean-speed that is obtained.

(c) It has been shown that

$$Q_1(t_w - t_1) = y_1$$

where Q_1 is the flow of vehicles with a speed of 100 km/h. In this case it is $0 \cdot 4 \times 2240$ or 896 veh/h,
t_w is the travel time of the observer, in this case, $0 \cdot 071$ h,
t_1 is the travel time of the vehicles travelling at 100 km/h, that is $0 \cdot 05$ h,
y_1 is the number of vehicles travelling at 100 km/h that overtake the observer.
Then

$$896(0 \cdot 071 - 0 \cdot 05) = y_1$$

$$= 18 \text{ or } 19 \text{ vehicles}$$

16
Intersections with priority control

Intersections are of the greatest importance in highway design because of their effect on the movement and safety of vehicular traffic flow. The actual place of intersection is determined by siting and design and the act of intersection by regulation and control of the traffic movement. At an intersection a vehicle transfers from the route on which it is travelling to another route, crossing any other traffic streams which flow between it and its destination. To perform this manoeuvre a vehicle may diverge from, merge with, or cross the paths of other vehicles.

Priority control of traffic at junctions is one of the most widely used ways of resolving the conflict between merging and crossing vehicles. The universal adoption of the 'Give Way to traffic on the right' rule at roundabouts together with the use of 'Give Way' and 'Stop' control at junctions has considerably increased the number of occasions at which a driver has to merge or cross a major road traffic stream making use of gaps or lags in one or more conflicting streams.

Both urban and rural motorways of the future will make use of priority control at grade separated junctions where merging vehicles have to enter a major traffic stream and only a few authorities plan to use traffic signal control for the junctions on their proposed inner ring roads. At the present time the relatively low volumes of flow during non-peak hours and the difficulty of dealing with a large proportion of right turning vehicles at traffic signals leads to the roundabout being preferred in many cases.

Accidents at priority intersections

Where these merging, diverging and crossing manoeuvres take place a potential, if not an actual, collision may take place between vehicles. The point of a potential or actual collision and the zone of influence around it has been defined as a conflict area.

The number and type of conflict areas may be taken as a measure of the accident potential of a junction, as shown in figures 16.1 and 16.2. It is shown in figure 16.1 that for a four-way priority type junction there are 16 potential crossing conflicts, 8 diverging conflicts and 8 merging conflicts. When this same intersection is converted to a right–left stagger as shown in figure 16.2, then there are only 6 crossing conflicts, 6 diverging conflicts and 6 merging conflicts. While there are objections to such a simplified approach to intersection safety, this method does give an indication of the increase in safety achieved by a staggered crossing.

Intersections may be divided into the following types: (a) priority intersections;

150

(b) signalised intersections; (c) gyratory or rotary intersections; (d) grade-separated intersections.

In the first three types the conflicts between vehicles are resolved by separation in time while in the fourth type separation in space is used to resolve the main vehicle

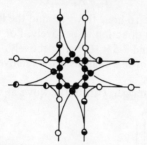

Figure 16.1 Conflict areas at a crossroads

● 16 crossing conflicts
○ 8 diverging conflicts
◐ 8 merging conflicts

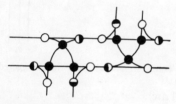

Figure 16.2 Conflict areas at a staggered crossroads

● 6 crossing conflicts
○ 6 diverging conflicts
◐ 6 merging conflicts

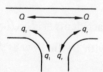

Figure 16.3 Vehicle streams at a three-way intersection

conflicts. In the selection of intersection type the following factors should be considered: traffic volumes and delay design speeds; pedestrian movements; cost and availability of land; the accident record of the existing intersection.

An interesting investigation into accidents at rural three-way junctions has been carried out by J. C. Tanner of the Transport and Road Research Laboratory. Traffic flows entering or leaving the minor road were divided into flows around the left and right-hand shoulders of the junction respectively as shown in figure 16.3 making the assumption that flows in opposite directions in the same paths were equal.

By means of regression analysis of accident data obtained from 232 rural junctions it was found that an approximate relationship between the annual accident rate around the left and right-hand shoulders of the junction A_l and A_r respectively could be given by the equations

$$A_l = R_l\sqrt{q_1 Q}$$
$$A_r = R_r\sqrt{q_r Q}$$

where q_1, q_r, Q are the August 16-hour flows around the left and right-hand shoulders and on the major road in each direction respectively. For the data studied R_r and R_l were found to be $4\cdot5 \times 10^{-4}$ and $7\cdot5 \times 10^{-4}$, respectively.

An interesting application of this analysis is the relative safety of left–right and right–left staggered crossroads

Considering the left–right stagger with traffic flows as shown in figure 16.4

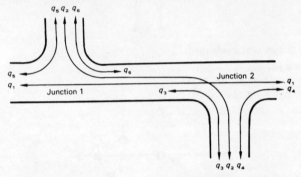

Figure 16.4 Traffic streams at a left–right staggered crossroads

then accidents at junction 1 are

$$A_1 = R_r\sqrt{(q_5 q_1)} + R_l\sqrt{(q_1(q_6 + q_2))}$$

and accidents at junction 2 are

$$A_2 = R_l\sqrt{(q_1(q_2 + q_3))} + R_r\sqrt{(q_4 q_1)}$$

Neglecting all values $< q_1 q_2$

$$\text{Total accidents} = 2R_l\sqrt{(q_1 q_2)}$$

Considering the right–left stagger with traffic flow as shown in figure 16.5, accidents

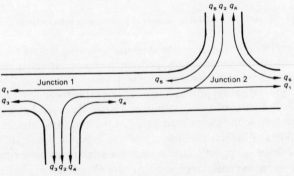

Figure 16.5 Traffic streams at a right–left staggered crossroads

at junction 1 are

$$A_1 = R_r\sqrt{(q_2 + q_4)q_1} + R_1\sqrt{(q_3q_1)}$$

and accidents at junction 2 are

$$A_2 = R_r\sqrt{(q_5 + q_2)q_1} + R_1\sqrt{(q_6q_1)}$$

Neglecting all values $< q_1q_2$

$$\text{Total accidents} = 2R_r\sqrt{(q_1q_2)}$$

Therefore, ratio of accidents $L/R : R/L$

$$= R_1 : R_r$$

$$= 7\cdot5 : 4\cdot5$$

For this reason the R/L stagger is usually considered safer.

A further application of this statistical relationship between accidents around the right and left-hand shoulders is the reduction in accidents that may be obtained by restriction of access to a major traffic route.

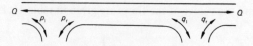

Figure 16.6 Traffic flows at the junction of two minor roads with a major road

Consider the geometric layout and the traffic flows shown in figure 16.6. Total accident potential is

$$R_r\sqrt{(p_rQ)} + R_r\sqrt{(q_rQ)} + R_1\sqrt{(p_2Q)} + R_1\sqrt{(q_2Q)}$$

$$= R_r\sqrt{Q}(\sqrt{p_r} + \sqrt{q_r}) + R_1\sqrt{Q}(\sqrt{q_1} + \sqrt{p_1})$$

$$= \sqrt{Q}[R_r(\sqrt{p_r} + \sqrt{q_r}) + R_1(\sqrt{q_1} + \sqrt{p_1})]$$

Consider now the alternative geometric layout shown in figure 16.7 but with the same traffic flows. Accidents at the junctions of the minor roads are ignored because they are given by the product of small quantities.

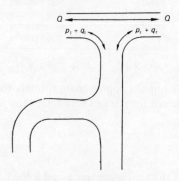

Figure 16.7 Traffic flows at the junction of the combined minor roads with a major road

Total accident potential is

$$R_1\sqrt{(p_1 + q_1)}Q + R_r\sqrt{(p_r + q_r)}Q = \sqrt{Q}[R_1\sqrt{(p_1 + q_1)} + R_r\sqrt{(p_r + q_r)}]$$

The ratio of accident potential for the two geometric layouts is then

$$R_r(\sqrt{p_r} + \sqrt{q_r}) + R_2 (\sqrt{q_1} + \sqrt{p_1}) : R_2\sqrt{(p_1 + q_1)} + R_r\sqrt{(p_r + q_r)}$$

Since $\sqrt{a} + \sqrt{b} > \sqrt{(a + b)}$ the accidents with the first layout will thus be greater than with the second layout. Greatest reduction occurs when $p_r = q_r$ and $p_1 = q_1$. The ratio of accidents at the two types of junction is then

$$(2\sqrt{p_r} + 2\sqrt{p_1}) : \sqrt{2}p_1 + \sqrt{2}p_r = 2(\sqrt{p_r} + \sqrt{p_1}) : \sqrt{2}(\sqrt{p_1} + \sqrt{p_r})$$

$$= 2 : 1{\cdot}41$$

$$= 1 : 0{\cdot}71$$

The maximum reduction is therefore approximately 30 per cent.

Problems

Annual traffic flows at a crossroad junction under priority control are illustrated diagrammatically and the values of the flows are given in table 16.1.

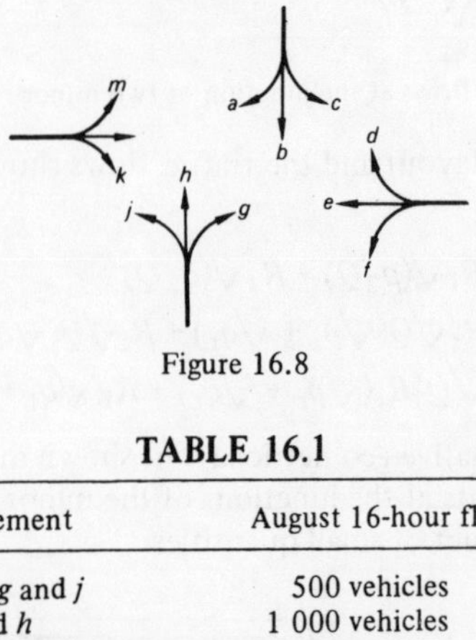

Figure 16.8

TABLE 16.1

Movement	August 16-hour flow
a, c, g and j	500 vehicles
b and h	1 000 vehicles
d, f, k and m	600 vehicles
e and l	10 000 vehicles

$$R_r = 4{\cdot}5 \times 10^{-4} \qquad R_1 = 7{\cdot}5 \times 10^{-4}$$

Is the error involved in using the approximate formula for the total accidents when a right–left stagger is employed for streams b and h greater or less than 25 per cent of the true value?

Solutions

With a right–left stagger of the traffic streams b and h the flow diagram is as shown in figure 16.9.

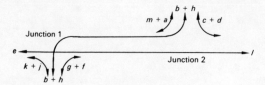

Figure 16.9

Accidents at junction 1 are

$$A_1 = \frac{R_r}{2} \sqrt{(b + h + g + f)(l + e)} + \frac{R_1}{2} \sqrt{(k + j)(l + e)}$$

Accidents at junction 2 are

$$A_2 = \frac{R_r}{2} \sqrt{(m + a + b + h)(l + e)} + \frac{R_1}{2} \sqrt{(c + d)(l + e)}$$

Total accidents at intersection are

$$A = \frac{R_r}{2} \sqrt{(b + b + g + f + m + a + b + h)(l + e)} + \frac{R_1}{2} \sqrt{(k + j + c + d)(l + e)}$$

$$= \frac{R_r}{2} \sqrt{124\,000\,000} + R_1 \sqrt{44\,000\,000}$$

$$= 5565 R_r + 3315 R_1$$

$$= 2 \cdot 5 + 2 \cdot 3$$

$$= 4 \cdot 8 \text{ accidents}$$

Ignoring the product of small quantities

$$A = R_r \sqrt{(l + e)(b + h)}$$
$$= R_r \sqrt{40\,000\,000}$$
$$= 2 \cdot 9 \text{ accidents}$$

Therefore the error involved in using the approximate formula is greater than 25 per cent of the true value.

17

Driver reactions at priority intersections

Interaction between traffic streams is an important aspect of highway traffic flow. It occurs when a driver changes traffic lane, merging with or crossing a traffic stream. Probably it takes place most frequently when priority control is used to resolve vehicular conflicts at highway intersections.

When a minor road driver arrives at an intersection he may either enter the major road by driving into a gap in the major road traffic stream or he may reject it as being too small and wait for a subsequent gap. It is important to differentiate when taking observations between lags and gaps. A lag is an unexpired portion of a gap which remains when a minor road driver arrives at the junction of the major and minor roads.

A minor road driver may make only one acceptance decision but he may make many rejections. If every decision of each driver is included in the observations then the resulting gap or lag acceptance distribution will be biased towards the slower driver.

To avoid bias in the observations it is possible to observe only first driver decisions. In this way most of the observed values will be lags and acceptances will be made by drivers from a rolling start. Second, third and subsequent driver decisions will be gap acceptances and will normally be higher in value because of the stationary start.

The usual hypothesis of minor road driver behaviour at priority intersections is the time hypothesis. It is assumed that a gap and lag acceptance judgment is formed as if a minor road driver made a precise estimate of the arrival time of the approaching major road vehicle. The minor road driver is thus able to estimate the time remaining before the approaching vehicle reaches the intersection and merges or crosses on the basis of the time remaining.

Observation of gap and lag acceptance

Lag acceptance behaviour can be studied by observing driver reactions at a priority intersection where the flow on the minor road exceeds 100 veh/h and where the major road traffic stream or streams exceed 400 veh/h.

Observations should be taken of one class of minor road vehicle making one type of turning movement, preferably observing left-turning cars to reduce the time required to complete the observations.

156

Using a stopwatch the time is noted at which a minor road vehicle arrives at the intersection and also the time at which the next conflicting major road vehicle arrives at the intersection. It is noted also whether the minor road driver accepts the lag and drives out into the major road or rejects the lag and remains in the minor road. After the first driver decision all further actions of the driver are ignored to reduce bias in the observations.

The difference between the two recorded times is the accepted or rejected lag. When observations are taken with a stopwatch it is normally only possible to record lag acceptance to the nearest 0·5 second and so it is convenient to group the observed lag acceptances and rejections into one-second classes.

A set of observations for left-turning passenger cars is shown in table 17.1. Notice that observation was continued until at least 25 observations were obtained in each class.

Column 1 contains the classes that run between zero and 100 per cent acceptance. Columns 2 and 3 give the numbers of observed lags that were rejected and accepted in each class. Column 4 is obtained by dividing column 3 by the sum of columns 2 and 3, the resulting value being expressed as a percentage.

TABLE 17.1   Observed rejected and accepted lags for left-turning vehicles

1	2	3	4
Lag class (s)	Number of observed rejections	Number of observed acceptances	Percentage observed acceptance
0·5–1·4	25	0	0
1·5–2·4	82	2	2
2·5–3·4	56	23	29
3·5–4·4	47	23	33
4·5–5·4	28	41	59
5·5–6·4	17	41	71
6·5–7·4	5	30	86
7·5–8·4	2	48	96
8·5–9·4	0	30	100

If the percentage acceptance is plotted against lag class mark as in figure 17.1 then it can be seen that the curve is approximately of cumulative normal form. This is to be expected in any action that is dependent on human reaction.

It is often necessary in theoretical traffic flow studies and in simulation work to find a mathematical distribution that approximates to an observed phenomenon.

When lag acceptance is of a cumulative normal form then the distribution may be written

Proportion of drivers accepting a lag of t second $= 100/(\sigma\sqrt{2\pi}) \int_{-\infty}^{t} \exp\left(-(t - \bar{t})^2/2\sigma^2\right) \, dt$

where $\bar{t}$ and σ are the mean and standard deviation.

A demonstration of the fit of the observed lag acceptance to a normal distribution may be obtained by plotting the probit of the acceptance against the lag class mark. Values of probits may be obtained from reference 3 (see page 143) or can be read from figure 17.2. The use of this technique converts a cumulative normal curve into a straight

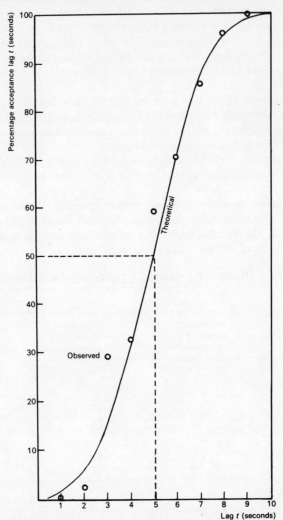

Figure 17.1 Lag acceptance distribution at a priority intersection

line and allows the mean and the standard deviation and also the theoretical values of acceptance to be estimated.

In figure 17.2 the best straight line is drawn through the observed point values. The mean lag acceptance t is the value when the probit is 5 and the standard deviation of lag acceptance is equal to minus the reciprocal of the slope of the line. From figure 17.2 these values are

$$\bar{t} = 5\cdot0 \text{ seconds}$$

$$\sigma = 1\cdot8 \text{ seconds}$$

The theoretical percentage acceptances can also be read from the graph and are given in table 17.2. The theoretical lag acceptance curve and the observed values are shown in

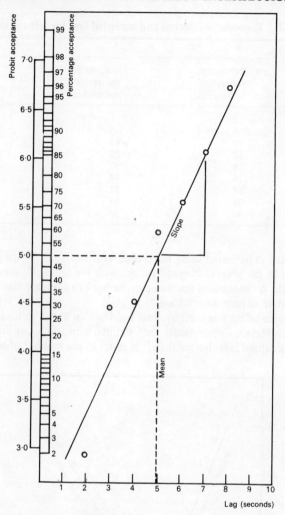

Figure 17.2 Transformation of cumulative normal distribution to a linear relationship using probit acceptance

figure 17.1 but a statistical test of the closeness of fit can be made by the chi-squared test. Column 8 is calculated from columns 3 and 7. The first three rows have been summed to give an adequate number of acceptances in each group.

The summed value of chi-squared has four degrees of freedom: the seven groups less the three constraints; the mean and standard deviation of the lag acceptances; and the total number of decisions.

From statistical tables it can be seen there is not a significant difference at the 5 per cent level between the observed and theoretical distribution.

Many observations of merging performance of vehicles have indicated that a cumulative log-normal distribution of lag acceptance is a better fit than the cumulative normal

TABLE 17.2   Theoretical rejected and accepted lags for left-turning vehicles

5	6	7	8
Lag class (seconds)	Percentage theoretical acceptance	Number of theoretical acceptances	Chi-squared
0·5–1·4	2	1	
1·5–2·4	6	5	
2·5–3·4	16	13	1·90
3·5–4·4	32	22	0·05
4·5–5·4	54	37	0·43
5·5–6·4	74	43	0·09
6·5–7·4	88	31	0·03
7·5–8·4	96	48	0
8·5–9·4	99	30	0

$$\Sigma\ 2·50$$

distribution. This can be tested using the probit transformation by the graphical fitting of a straight line to the observed lag acceptance with the log lag class-mark replacing the lag class-mark. A chi-squared test can then be used to compare the closeness of fit with the cumulative normal distribution.

A different value of lag was used by Raff and Hart[1] in investigations into driver behaviour in New Haven, Connecticut. They defined a critical lag as that lag of which the number of accepted lags shorter than it is equal to the number of rejected lags longer than it.

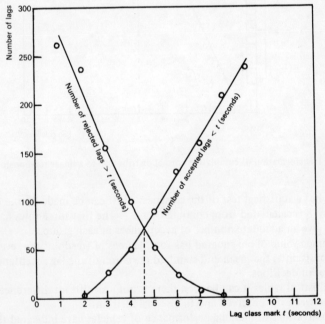

Figure 17.3 Graphical determination of the critical lag

This critical lag is obtained graphically in figure 17.3 where the critical lag is determined graphically to be 4·6 seconds. This approach is only useful in giving a value of lag acceptance that may be used in traffic studies that use average values of driver performance. It was used by Tsongos and Weiner[2] to compare the effect of darkness on gap acceptance, a subject of considerable interest because of the proportion of peak hour flows that occurs during the hours of darkness.

It has been shown that the critical lag and the mean of the lag acceptance distribution are related in the case where the lag acceptance distribution is normally distributed. This relationship is

$$\text{critical lag} = \bar{t} - \sigma^2 q/2$$

where $\bar{t}$ and σ are the mean and standard deviation of the lag acceptance distribution and q is the flow of vehicles on the major road.

References

1. M. S. Raff and J. W. Hart. A volume warrant for urban stop signs. The Eno Foundation for Highway Traffic Control, Saugatuck, Conn., U.S.A. (1950).
2. N. G. Tsongos and S. Weiner. Comparison of day and night gap acceptance probabilities. *Publ. Rds, Wash.* 35 (1969), 7, 157–65.

Problems

1. Observations of driver lag acceptance at an intersection showed that it was normally distributed with a mean of 5·0 s and a standard deviation of 1·5 s.
(a) Is the percentage of drivers requiring a lag greater than 7·0 s, 9 per cent, 15 per cent or 31 per cent?
(b) If it is noted that the mean of the combined lag and gap acceptance distribution is 6·0 s would this indicate

(i) that drivers entering the major road without stopping are likely to enter the major road more readily than those drivers who have come to a halt before entering?
(ii) that the second decision gap acceptance of drivers has a lower mean value than first decisions?
(iii) neither of these?

(c) What is the value of the critical lag at this intersection when the major road flow is 600 veh/h?

Solutions

1. (a) The percentage of drivers accepting a lag greater than t seconds may be obtained from the straight-line relationship between the probit of the percentage of lag acceptance and the time lag, which may be stated as

$$\text{probit percentage acceptance} = a + b \times \text{lag } t \text{ s}$$

where b = the reciprocal of the standard deviation of lag acceptance. From the value of the mean and standard deviations given and probit tables

$$5 = a + \frac{5}{1 \cdot 5}$$

and

$$a = 1 \cdot 67$$

so that

$$\text{probit percentage acceptance} = 1 \cdot 67 + \frac{\text{lag}}{1 \cdot 5}$$

When the lag is 7·0 s

$$\text{probit percentage acceptance} = 1 \cdot 67 + \frac{7 \cdot 0}{1 \cdot 5}$$

$$= 6 \cdot 34$$

From probit tables (or from figure 17.2) percentage acceptance = 91 per cent. The percentage of drivers who require a lag greater than 7·0 s is therefore

$$100 - 91 \text{ per cent} = 9 \text{ per cent}$$

(b) As the combined lag and gap acceptance distribution has a mean of 6·0 s and the lag acceptance distribution has a mean of 5·0 s this indicates that the gap acceptance distribution has a larger value of mean acceptance than the lag acceptance. Hence (1) is correct since drivers who enter the major road without stopping and hence accept lags do so more readily than those who come to a halt.

(c) From the relationship

$$\text{critical lag} = \bar{t} - \sigma^2 q/2$$

$$\bar{t} = 5 \cdot 0 \text{ s}$$

$$\sigma = 1 \cdot 5 \text{ s}$$

$$q = 600/3600 \text{ veh/s}$$

$$\text{critical lag} = 5 \cdot 0 - 1 \cdot 5^2/6 . 2$$

$$= 4 \cdot 8 \text{ s}$$

18

Delays at priority intersections

In contrast to traffic signal-controlled junctions, where saturation flows and resulting junction capacities can be easily calculated, the estimation of the practical capacity of priority type junctions presents considerable difficulties.

At the present time the growth of traffic on a largely unimproved urban highway system has resulted, even in the smaller urban areas, in considerable congestion and delay at intersections during peak hours. These symptoms of traffic growth were reached many years ago in the United States and it is not surprising that some of the first work on the capacity of priority type junctions was carried out in the U.S.A. A measure of the practical capacity of an intersection is the average delay to minor road vehicles. At volumes beyond the practical capacity the average delay increases considerably with only small increases in volume.

Morton S. Raff of the Bureau of Highway Traffic, Yale University and Jack W. Hart[1] of the Eno Foundation for Highway Traffic Control made observations on the delays to vehicles at four intersections in the built-up areas of New Haven, Connecticut. They hoped to obtain an empirical equation connecting minor and major road volumes with average delays to minor road vehicles. Observations were made using a multi-pen recorder to note the time at which vehicles passed through the intersection and it was possible by this means to note traffic volumes and delays.

Typical of their observations is figure 18.1, which shows the average wait for all minor road vehicles. The considerable scatter of the results could not be explained by the investigators and they were, not surprisingly, unable to obtain an empirical relationship connecting main road volume and average delay.

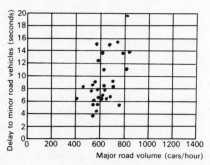

Figure 18.1 Observed variations in delay to minor road vehicles (adapted from ref. 1)

An analytical approach to delay

It was not until 1962 that a mathematical formula or model was proposed which connected the many variables inherent in calculating the average delay to side road vehicles. J. C. Tanner[2] proposed a formula for the average delay $\bar{w}_2$ to minor road vehicles.

The variables used to calculate $\bar{w}_2$ in this formula are

1. The major road volume q_1.
2. The minor road volume q_2.
3. The minimum time headway between major road vehicles β_1.
4. The minimum time headway between minor road vehicles emerging from the minor road β_2.
5. The average lag or gap α in the major road traffic stream accepted by minor road drivers when entering the major road traffic stream.

These variables and their determination for a particular junction will now be explained.

No explanation of the calculation of q_1 and q_2 is required but it should be noted that the formula for delay was propounded for a T-junction where the merging or crossing movement is into or through a single unidirectional traffic stream. In the formula these volumes are normally substituted as vehicles per second.

The introduction of a minimum time headway β_1 between major road traffic introduces an element of bunching or platooning into the flow caused by the inability of vehicles to overtake. Time headway is the interval of time between the fronts of successive vehicles passing a point on the highway. The usual method of estimating β_1 is to note the time headway between the front and rear vehicle of a bunch or platoon of vehicles which are following each other at their minimum separation. The average time headway between them is then calculated. The process is repeated for a considerable number of bunches until an average value representative of the traffic flow is obtained.

The minimum time headway β_2 between minor road vehicles emerging into the major road represents the time between successive minor road drivers who perceive the gap or lag available to them in the major road flow and follow each other into the major road. It is estimated by observing the average time headway between minor road vehicles following each other into the major road.

The last variable needed to estimate the average delay at a priority intersection is the average gap or lag α in the major road stream that is accepted by the minor road driver.

Observations are made of the time at which a minor road vehicle arrives at the junction with the major road and also of the time at which the next major road vehicle arrives at the junction. The gap or lag is the difference between these two times. It is also observed whether the minor road driver enters the first gap or lag that is available to him in the major road traffic stream. If he enters the major road traffic stream, then he is said to have accepted the first available gap or lag, while if he waits until a subsequent gap to enter the major road traffic he is said to have rejected the first available gap or lag. To avoid bias in favour of slower drivers, once a driver has rejected a gap his subsequent performance is neglected.

A lag is an unexpired portion of a gap that is presented to a minor road driver arriving at the junction after the commencement of a gap.

Gaps and lags are classified into classes of suitable size and the percentage of acceptance for each class calculated. When the percentage acceptance is plotted against the gap or lag class mark the curve obtained is of a cumulative normal form and it is possible by statistical means to estimate the most likely value of the 50 percentile acceptance.

With these variables known it is possible to estimate the average delay to minor road ehicles at varying major and minor road traffic volumes using the formula

$$\bar{w}_2 = \frac{\frac{1}{2}E(y^2)/Y + q_2 Y \exp(-\beta_2 q_1)[\exp(\beta_2 q_1) - \beta_2 q_1 - 1]/q_1}{1 - q_2 Y[1 - \exp(-\beta_2 q_1)]}$$

$$E(y) = \frac{\exp[q_1(\alpha - \beta_1)]}{q_1(1 - \beta_1 q_1)} - \frac{1}{q_1}$$

$$E(y^2) = \frac{2 \exp[q_1(\alpha - \beta_1)]}{q_1^2(1 - \beta_1 q_1)^2} \{\exp[q_1(\alpha - \beta_1)] - \alpha q_1(1 - \beta_1 q_1) - 1$$

$$+ \beta_1 q_1 - \beta_1^2 q_1^2 + \frac{1}{2}\beta_1^2 q_1^2/(1 - \beta_1 q_1)\}$$

$$Y = E(y) + 1/q_1$$

In addition Tanner derived an expression for the maximum discharge from a minor road and this is given below, where the symbols have the same meaning as before.

$$q_2(\text{max}) = \frac{q_1(1 - \beta_1 q_1)}{\exp[q_1(\alpha - \beta_1)][1 - \exp(-\beta_2 q_1)]}$$

An indication of likely values of these junction flow parameters can be obtained from information that was given in a previous Department of Transport junction capacity assessment method when β_1 was taken as 2 s for minor road vehicles that merge with or cross one major traffic stream and as 1 s when they cross two major road traffic streams. β_2 is taken as 3 s and the value of α varies between 4 s and 12 s according to intersection layout and design speed.

The interactive approach to capacity

Whilst the previously discussed gap acceptance approach describes an important feature of the operation of priority intersections there are a number of difficulties in the use of these methods for estimating the capacity of a junction. Accurate measurement of the parameters that influence capacity are not easy to measure and so there are inaccuracies in the estimation of capacity. Also under heavily-trafficked conditions some major road vehicles give way to the more aggressive minor road drivers and gap acceptance operation breaks down.

An investigation into the relationship between the capacity of non-priority flows and priority flows at major/minor junctions has been carried out by the Transport and Road Research Laboratory and Martin and Voorhees Associates[3]. Subsequently the results of the work were incorporated into a Department of Transport advice note on the design of major/minor junctions[4].

In this approach to design, the capacity of a junction is taken to be the traffic situation when there is a continuous queue feeding one or more turning movements at the junction. Not all the movements need be in this state for the junction to be considered to be at capacity.

Consider the traffic streams shown in figure 18.3 when a major road has arms A and C and a minor road is denoted by arm B. The traffic flows at the junction are represented by q_{c-A} etc. as shown and the streams for which it is required to calculate capacity are

$$q_{b-c}, q_{b-a}, q_{c-b}$$

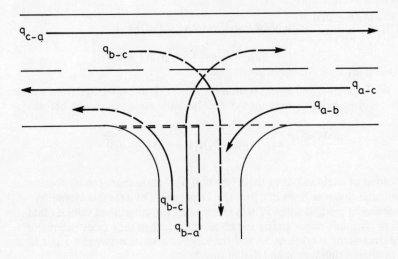

Figure 18.3

In this interactive approach the capacities of these streams depends on the flows of one or more of the priority streams and also on the geometric feature of the junction.

The best predictive equations for turning stream capacity (sat. q) when the major road has approach speeds not exceeding 85 kph are given as:

$$\text{sat. } q_{b-a} = D(627 + 14W_{cr} - Y[0.364q_{a-c} + 0.114q_{a-b} \\ + 0.229q_{c-a} + 0.520q_{c-b} \tag{18.3}$$

$$\text{sat. } q_{b-c} = E(745 - Y[0.364q_{a-c} + 0.144q_{a-b}]) \tag{18.4}$$

$$\text{sat. } q_{c-b} = F(745 - 0.364Y[q_{a-c} + q_{a-b}]) \tag{18.5}$$

where $Y = (1 - 0.0345W)$ and W is the width of the major road (in metres).

In the above equations for the saturated capacities the geometric characteristics of the junction are represented by D, E and F and are stream-specific:

$$D = [1 + 0.094(wb - a - 3.65)] \ [1 + 0.0009(Vrb - a - 120)]$$
$$[1 + 0.0006(Vlb - a - 150)]$$

$$E = [1 + 0.094(wb - c - 3.65)] \ [1 + 0.0009(Vrb - c - 120)]$$

$$F = [1 + 0.094(wc - b - 3.65)] \ [1 + 0.0009(\dot{V}rc - b - 120)]$$

where $wb - a$ denotes the average lane width over a distance of 20 m available to waiting vehicles in the stream $b - a$, and $Vrb - a$, $Vlb - a$ are the corresponding visibilities to the right and left measured in the normal manner for visibility splays. In these equations capacities and flows are in pcu/hour where a heavy goods vehicle is equivalent to 2 pcu. Detailed information on the measurement of geometric characteristics is given in references 3 and 4.

Queues and delays at oversaturated junctions

In the study of junctions it is frequently important to determine queue lengths and delays. When demand is less than capacity then delay and consequently queue length can be predicted using a steady-state approach such as that theorised by Tanner[2]. Using steady-state theory, delays increase as demand approaches capacity and become infinite when demand reaches capacity. In the real-life situation when demand is close to capacity or even when the capacity is exceeded for short periods, the queue length growth lags behind that predicted by steady-state theory.

Deterministic queueing theory in which the number of vehicles delayed is the difference between capacity and demand ignores the statistical nature of vehicle arrivals and departures and seriously underestimates delay. Using deterministic theory, delay is zero when demand equals capacity.

In many traffic situations demand is close to capacity and even exceeds it for short periods of time and a combination of both steady-state and deterministic theory has been proposed by Kimber and Hollis[5]. Using a coordinate transformation technique they developed the following expressions for queue length and delay:

$$\text{Queue length} = \frac{1}{2}((A^2 + B)^{\frac{1}{2}} - A \tag{18.6}$$

where

$$A = \frac{(1 - \rho)(\mu t)^2 + (1 - L_0)\mu t - 2(1 - C)(L_0 + \rho\mu t)}{\mu t + (1 - C)}$$

$$B = \frac{4(L_0 + \rho\mu t)[\mu t - (1 - C)(L_0 + \rho\mu t)]}{\mu t + (1 - C)}$$

$$\text{Delay per unit time} = \frac{1}{2}((F^2 + G)^{\frac{1}{2}} - F \tag{18.7}$$

where

$$F = \frac{(1 - \rho)(\mu t)^2 - 2(L_0 - 1)\mu t - 4(1 - C)(L_0 + \rho\mu t)}{2(\mu t + 2(1 - C))}$$

$$G = \frac{2(2L_0 + \rho\mu t)\left[\mu t - (1 - C)(2L_0 + \rho\mu t)\right]}{\mu t + 2(1 - C)}$$

where
C = 1 for random arrivals and service,
C = 0 for regular arrivals and service,
μ = capacity,
ρ = q/μ,
q = demand,
t = time.

Demand on both major and minor roads increases during peak periods and then decreases, resulting in a decrease and an increase in capacity.

For any non-priority stream it is possible to calculate capacity by taking into account the levels of flow in the other interacting traffic streams and the geometric characteristics of the junction. If the variation of flow with time is known then the changes in queue length and delay can be calculated.

The computer program PICADY has been developed by the Traffic Engineering Department of the Transport and Road Research Laboratory to determine capacities, queues and delays at 3-arm major/minor priority junctions. It can cater for a single or dual carriageway major road, with or without a right-turning lane, and with a one or two lane approach on the minor road. The program models the build-up and decay of queues and delays through a peak period which can be examined at discrete time intervals through the period, normally at 5 to 15 minute intervals as selected by the user.

An example of the manual computation of queue length and delay using the results of interactive theory follows. At a junction of the type shown in figure 18.3 the demand flows shown in table 18.1 (in pcu) are noted and the capacity is then calculated from equation 18.4.

TABLE 18.1

Time	Demand b−c	Demand a−c	Capacity b−c
0815	200	900	508
0820	500	1000	481
0825	300	1500	350
0830	300	1400	376
0835	200	1100	455
0840	200	900	508

For each time interval the values of q, μ and ρ are calculated and given in table 18.2 and the queue length and delay calculated from equations 18.6 and 18.7.

TABLE 18.2

Time interval commencing	q_{b-c} (pcu/s)	μ_{b-c} (pcu/s)	ρ_{b-c}	Queue length at end of interval	Delay per unit time (S)
0800	0.056	0.141	0.397	0	1.6
0815	0.139	0.134	1.037	40.5	13.0
0830	0.083	0.097	0.856	34.3	26.4
0845	0.083	0.104	0.798	13.6	11.5
0900	0.056	0.126	0.444	1.70	1.0
0915	0.056	0.141	0.397		0.7

In the calculation of queue length and delay it was assumed that the queue was zero at 0800 hours and that vehicles arrived and departed at random. The calculated values are entered in table 18.2.

In the design of junctions the Department of Transport recommend[4] that a Design Reference Flow be selected to allow the detailed design of alternative feasible junctions to be carried out, the performance of which can then be assessed. When a peak hourly flow is selected to represent the Design Reference Flow the traffic function of the road should be considered. If a junction is designed to be adequate for the highest peak hour flow in a future design year then it is likely that the design will not be economically viable.

If the road has a recreational facility where peak summer flows are very much higher than at other times of the year, then it is suggested that the 200th highest hourly flow might be appropriate. On the other hand on an urban road where there are not likely to be very great variations in flow then the 30th highest flow is suggested as being suitable. The 50th highest flow is considered to be most appropriate for inter-urban roads.

References

1. M. S. Raff and J. W. Hart, A volume warrant for urban stop signs, The Eno Foundation for Highway Traffic Control, Saugatuck, Connecticut, U.S.A. (1950)
2. J. C. Tanner, A theoretical analysis of delays at an uncontrolled intersection, *Biometrika* **49** (1962), 163–70
3. R. M. Kimber and R. D. Coombe, The traffic capacity of major/minor priority junctions, Transport and Road Research Laboratory Supplementary Report 582, Crowthorne (1980)
4. Department of Transport, Departmental Advice Note TA 23/81, Junctions and Accesses: Determination of Size of Roundabouts and Major/Minor Junctions (1981)
5. R. M. Kimber and Erica M. Hollis, Traffic queues and delays at road junctions, Transport and Road Research Laboratory Report 909, Crowthorne (1979)

Solutions

1. The five traffic-flow characteristics used to calculate the theoretical delay to minor road vehicles at a priority intersection are

q_1 the major road flow with which minor road vehicles conflict,
q_2 the minor road flow,
β_1 the mean minimum headway between bunched vehicles on the major road,
β_2 the mean minimum headway between vehicles emerging from the minor road,
α the mean gap or lag accepted by minor road drivers.

2. The value of α inserted in the delay formula as derived by Tanner is the mean gap or lag acceptance and it would be appropriate to use the 50 percentile value when it approximates to a cumulative normal form.

3. (a) This statement is incorrect. It has been noted that even when it is assumed that drivers make gap and lag acceptance decisions on the basis of the time hypothesis, drivers entering high speed roads require considerably larger gaps and lags than when speeds on the major road are lower.

(b) This statement is correct. The assumption made by Tanner when deriving the delay to minor road vehicles at a priority intersection was that a single minor road stream conflicted with a single major road stream. These assumptions are correct for the situation in which a single stream of minor road vehicles turn right into a single major road stream.

(c) This statement is incorrect. The assumption made by Tanner in deriving the delay expression was that major road vehicles arrived at random at the intersection but could not pass through it at a smaller time headway than β_1. Bunching of major road vehicles is therefore likely in heavily trafficked situations.

19

A simulation approach to delays at priority intersections

Prediction of capacities and delays at highway intersections under priority control may be made by the use of empirical, mathematical or simulation models.

The dependent variable in these models is usually the average delay to minor road vehicles and measures the effectiveness of the junction in allowing conflicting traffic streams to merge or intersect. The performance of the junction may be measured and alternative geometric designs evaluated by consideration of average delays to minor road vehicles.

Empirical models attempt to predict highway capacity on the basis of past observations and probably the best known example of this approach is the Highway Capacity Manual[1] calibrated by traffic data collected throughout the United States. Other researchers[2-5] have studied traffic flow at priority type intersections and used regression analysis methods to obtain relationships between average delay and traffic volumes.

At the present time the average delay to minor road vehicles and the maximum discharge from the minor road at a priority intersection are calculated using mathematical models given by Tanner[6].

Both mathematical and empirical models have limitations in their use. Traffic flow at an intersection is such a complex phenomenon that even the most complex mathematical model has to adopt a macroscopic approach in that all vehicles and drivers generally have the same characteristics.

An empirical model would require a considerable amount of data for its calibration and it is often very difficult to obtain this data with the precision, in the quantity or at the required traffic volume levels. Often the traffic volumes required may, when found, not persist long enough for sufficient observations to be obtained.

If the effect of geometric features is being investigated it will also be necessary for a junction having these features to be located or else constructed in the field or on a test track before traffic observations can be obtained.

While the development and verification of a simulation model is a complex process it is not subject to many of the disadvantages of empirical and mathematical models. Simulation models of intersection traffic flow can be flexible enough to cover a wide range of highway and traffic conditions. Inputs to the model can be specified to any distribution and the form of traffic control and driver characteristics varied. Simulation time may be as long as desired and several figures of merit may be printed out during the simulation process.

A simulation model requires the formation of a model system which represents the real situation at the site being studied. Two forms of simulation have been used in the study of highway traffic flow, analog simulation and digital simulation. In analog simulation an analogy is made between the real world situation and an analog physical system, the components of the analog physical system interacting in the same manner as in the real situation.

In digital simulation the state of the simulated system is stored in digital form and the situation is updated in accordance with stored instructions or rules of the model. Updating of the system may be carried out in two ways, by regular time scanning or by event scanning.

When regular time scanning is used the state of the traffic system is examined at equal scan intervals and all required vehicle movements calculated for the next interval in time. Updating of the traffic situation is carried out for all components of the model. The process is then repeated for the next time scan period.

The length of the time scan period selected will affect the precision of the model. It must be small enough to include all vehicle actions of significance but if the scan interval selected is too small the computer run time will be increased. The time scan interval selected will depend on the computer resources available, the objects of the simulation process and the method of generation of vehicular headways.

When event scanning is used the events that have the greatest importance in the operation of the program are used to build the program. The time at which the next significant event occurs is determined by the program and the traffic situation updated to that event.

To evolve a simulation program, it is usually most convenient to draw up a flow chart showing the steps in the simulation process. Figure 19.1 shows the flow chart for the simulation of the traffic system where a single stream of left-turning minor road vehicles intersect or merge with a single stream of major road vehicles. In this program regular time scanning is used and the figure of merit employed is the average delay to minor road drivers.

An appreciation of digital computer simulation can be obtained by a manual simulation. While this is laborious to carry out for any but a very limited real time period it illustrates the use that can be made of mathematical representations of headway and gap acceptance distributions in simulation.

As an illustration, the traffic flow at the intersection of a single stream of left-turning minor road vehicles and a stream of major road vehicles may be simulated. The major road flow headway distribution may be assumed to be represented by the double exponential distribution with the following parameters

$$\bar{t}_1 = 2 \cdot 5 \text{ s}$$

$$\bar{t}_2 = 5 \cdot 5 \text{ s}$$

$$L = 0 \cdot 5$$

where t_1, t_2 and L are the mean time headway of free-flowing vehicles, the mean time headway of restrained vehicles and the proportion of restrained vehicles in the traffic flow respectively.

The minor road flow headway-distribution may be assumed to be represented by the negative exponential distribution—but no vehicle may arrive with a time headway of

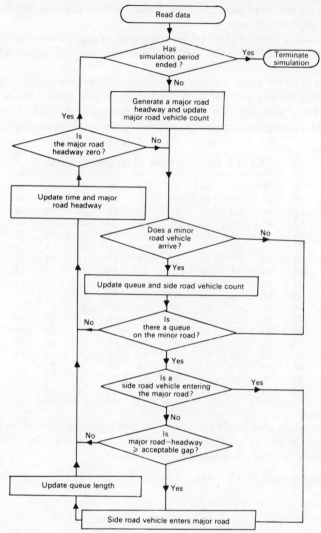

Figure 19.1 A flow chart for the simulation of a single minor road stream with a single major road stream

less than one second between it and the preceding vehicle. The single parameter in this distribution is

$$\bar{t} = 9 \text{ s}$$

where $\bar{t}$ is the mean headway. In addition minor road vehicles exit from the minor road at a minimum headway of $1 \cdot 0$ s.

Gap and lag acceptance for minor road drivers is assumed to be similar and to be represented by the normal distribution with the parameters

$$\bar{t} = 2 \cdot 5 \text{ s}$$

$$\sigma = 0 \cdot 9 \text{ s}$$

where $\bar{t}$ and σ are the mean gap and lag acceptance and the standard distribution of gap and lag acceptance respectively.

Furthermore it is noted that no drivers will accept or reject a lag or gap less than the mean minus one standard deviation or greater than the mean plus one standard deviation respectively.

Generation of major and minor stream headways will be carried out by Monte Carlo processes making use of a series of random numbers. Such a series is shown in table 19. and could be easily prepared from a series of ten identical cards, each carrying a number from 0 to 9, by picking cards at random and noting the numbers obtained.

TABLE 19.1   A table of random numbers

81	77	72	64	42	31	29	46	62	21
45	93	60	17	35	78	25	42	41	16
11	64	30	58	60	21	33	75	79	74
15	63	47	59	51	13	59	85	27	62
50	53	22	54	96	95	65	24	25	73
27	89	64	72	81	74	40	09	19	61
82	52	75	59	55	79	17	14	24	33
88	66	54	64	70	52	85	50	13	63
23	80	45	68	42	93	67	03	97	42
03	02	59	34	49	77	70	40	75	22
34	95	49	12	91	51	95	80	07	36
51	85	12	18	18	75	06	16	32	84
60	97	67	18	93	92	23	09	41	64
32	78	34	17	83	90	90	51	62	55
58	71	19	68	23	46	76	30	18	79
45	11	28	93	45	58	10	79	31	21
47	56	38	09	90	43	25	27	74	03
52	64	10	66	39	49	93	74	48	15

To generate major road headways, successive two digits are selected from table 19.1 and regarded as the first two decimal places of a random fraction. The random fraction may then be used to solve the following equation for t

$$(\text{random fraction}) = L \exp\left[-(t - e)/(\bar{t}_1 - e)\right] + (1 - L) \exp\left(-t/\bar{t}\right)$$

$$t \geqslant e$$

$$= L + (1 - L) \exp\left(-t/\bar{t}_2\right) \qquad (19.1)$$

$$0 < t < e$$

While the rapid solution of this equation to obtain t, the major road headway, would not present any difficulty when using a digital computer a graphical solution will be used in this example.

To obtain the major road headways by graphical means equation 19.1 is plotted as shown in figure 19.2. A series of two random numbers are then taken successively from table 19.1 and inserted in figure 19.2 to obtain the headways which are shown in table 19.2.

Successive headways on the minor road are next simulated by substituting successive random fractions in the cumulative negative exponential distribution

$$(\text{random fraction}) = \exp\left[-(t - 1)/(\bar{t} - 1)\right]$$

As with the major road headways, the successive solution of this equation will be made by a graphical method. The cumulative minor road headway distribution is shown in figure 19.3, and successive series of two random numbers are used as random fractions to generate headways as previously. The minor road headways generated in this way are given in table 19.2.

The third set of variables to be simulated are the lag and gap acceptances of minor road drivers. The same technique as has been used previously for generating headways will be used and successive random fractions are entered into the cumulative acceptance distribution as shown in figure 19.4. The acceptable lags obtained in this manner and

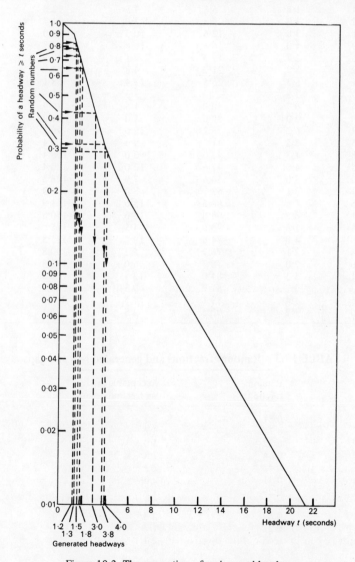

Figure 19.2 The generation of major road headways

TABLE 19.2 Generated major and minor road headways

Random fraction	Major road headway (seconds)	Cumulative major road headway (seconds)	Minor road headway (seconds)	Cumulative minor road headway (seconds)
81	1·2	1·2	2·6	2·6
77	1·3	2·5	3·1	5·7
72	1·5	4·0	3·6	9·3
64	1·8	5·8	4·5	13·8
42	3·0	8·8	7·9	21·7
31	3·8	12·6	10·3	32·0
29	4·0	16·6	10·8	42·8
46	2·7	19·3	7·1	49·9
62	1·9	21·2	4·8	54·7
21	5·3	26·5	13·4	68·1
45	2·8	29·3	7·3	75·9
93	0·7	30·0	1·5	76·9
60	2·0	32·0	5·0	81·9
17	6·2	38·2	15·0	96·1
35	3·5	41·7	9·3	105·4
78	1·3	43·0	3·0	108·4
25	4·6	47·6	12·0	120·4
42	3·0	50·6	7·9	128·3
41	3·1	53·7	8·0	136·3
16	6·6	60·3	15·6	151·9
11	8·5	68·8	18·6	170·5
64	1·8	70·6	4·5	175·0
30	4·0	74·6	10·5	185·5
58	2·1	76·7	5·3	190·8
60	2·0	78·7	5·0	195·8
21	5·3	84·0	13·4	209·2
33	3·7	87·7	9·8	219·0
75	1·4	89·1	3·3	222·3
79	1·2	90·3	2·8	225·1

TABLE 19.3 Random fractions and generated lags and gaps

Random fraction	Acceptable gap or lag (seconds)
0·81	3·3
0·77	3·2
0·72	3·0
0·64	2·8
0·42	2·3
0·31	2·0
0·29	1·9
0·46	2·4
0·62	2·8
0·21	1·7
0·45	2·4
0·93	3·8
0·60	2·7

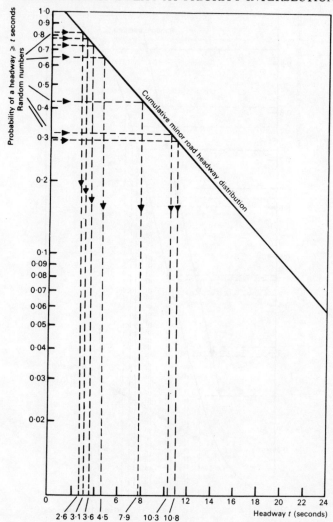

Figure 19.3 The generation of minor road headways

the successive random fractions used to generate them are given in table 19.3. It should be noted that the generated acceptable lags cannot exceed the mean lag plus two standard deviations; nor be less than the mean lag minus two standard deviations. Maximum and minimum values are thus 3·4 s and 1·6 s, respectively, and it is thus not necessary to amend any of the values generated.

All the variables required for the simulation of traffic flow have now been generated and it is possible to set up a table of events in which successive features of importance are arranged in order of their occurrence. This simulation table is shown as table 19.4 and initially the columns (1) and (2) are introduced from tables 19.2 and 19.3. Column (3) has the same value as column (2) if a queue does not exist. The value in column (3)

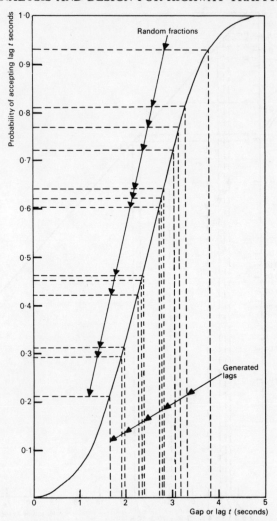

Figure 19.4 The generation of acceptable gaps and lags

is the lowest value of the time of arrival or the exit time of the previous vehicle in the queue plus one second.

Column (4), the required lag, is obtained from the previously generated lags in table 19.3 and the values are entered when a vehicle arrives at the stop or give way line.

A time value is entered in column (5) as soon as the gap or lag in the major road stream is equal to or greater than the required lag or gap. It is assumed that a vehicle leaves the minor road instantly but a further vehicle cannot leave the minor road for a further one second.

Column (6) records the queue length while column (7) is the difference between the time of arrival of a minor road vehicle and the time of exit. This column is summed and

TABLE 19.4 The simulation table

1	2	3	4	5	6	7
Major road vehicle arrives at junction	Minor road vehicle arrives at end of queue	Minor road vehicle arrives at stop or give way line	Required gap or lag (seconds)	Minor road vehicle enters major road	Queue length	Delay of existing vehicle (seconds)
1·2						
2·5						
	2·6	2·6	3·3	no	1	
4·0				no		
	5·7			no	2	
5·8				no	2	
8·8				8·8	1	6·2
		9·8	3·2	no	1	
	9·3			no	2	
12·6				12·6	1	6·9
		13·6	3·0	13·6	0	4·3
	13·8	14·6	2·8	no	1	
16·6				no	1	
19·3				no	1	
21·2				21·2	0	7·4
	21·7	22·2	2·3	22·2	0	0·5
26·5						
29·3						
30·0						
32·0						
	32·0	32·0	2·0	32·0	0	0
38·2						
41·7	42·8	42·8	1·9	no	1	

TABLE 19.4 continued

1	2	3	4	5	6	7
Major road vehicle arrives at junction	Minor road vehicle arrives at end of queue	Minor road vehicle arrives at stop or give way line	Required gap or lag (seconds)	Minor road vehicle enters major road	Queue length	Delay of existing vehicle (seconds)
43·0				43·0	0	0·2
47·6						
	49·9	49·9	2·4	no	1	
50·6				50·6	0	0·7
53·7						
	54·7	54·7	2·8	54·7	0	0
60·3						
	68·1	68·1	1·7	no	1	
68·8				68·8	0	0·7
70·6						
74·6						
	75·4	75·4	2·4	no	1	
	76·9			no	2	
76·7				no	2	
78·7				78·7	1	3·3
		79·7	3·8	79·7	0	2·8
	81·9	81·9	2·7	81·9	0	0
84·0						

cumulative delay 33·0

average delay = cumulative delay/number of minor road vehicles leaving intersection

= 33·0/13 s

= 2·5 s

when divided by the number of minor road vehicles entering the intersection gives the average delay to minor road vehicles.

For the traffic situation simulated the value of average delay is 2.4 s: the period of simulation is however extremely short and the simulation should be extended to a minimum period of 600 s. At high traffic flows the effect of assuming that a queue does not exist at the commencement of the simulation period tends to underestimate delays unless the simulation period is sufficiently long.

References

1. Highway Capacity Manual. *Highway Research Board Special Report* 87, 1965.
2. W. N. Volk. Effect of type of control on intersection delay. *Proc. Highw. Res. Bd* (1956), 523-33
3. DeLeuw, Cather and Co. The effect of regulatory devices on intersectional capacity and operations. Final Report Project 3-6. National Co-operative Highway Research Project, August (1966)
4. J. G. Wardrop and J. T. Duff. Factors affecting road capacity. International Study Week in Traffic Engineering, Stresa (1956)
5. F. V. Webster and B. M. Cobbe. Traffic signals. *Tech. Pap. Rd. Res. Bd* 56
6. J. C. Tanner. A theoretical analysis of delays at an uncontrolled intersection. *Biometrika*, 49 (1962), 163-70

20

Weaving action at intersections

Where traffic flows are small then the control of traffic movements at intersections may be achieved by priority control. As has been discussed previously the form of priority control in this country is that minor road vehicles give way to major road vehicles. On the continent of Europe nearside priority is used where vehicles give way to traffic approaching from the right, while in Australia off-side priority is used.

As traffic flows increase delays with priority control become excessive and at high levels of flow grade-separated junctions are necessary. A junction of this form is however extremely expensive: in addition, land requirements are great and in urban and suburban areas interference with pedestrian flow can be considerable. For these reasons at-grade intersections either of the roundabout or signal control type are extremely important in urban areas. The characteristics of these junction types have been described by Millard[1] and a consideration of these characteristics will usually determine which type of junction is appropriate. They include:

(a) In urban situations land requirements are usually the deciding factor. If this is so, it will be found that the land required for a large island roundabout is greater than that needed for traffic signal control. This is especially true if flows on one pair of arms are low. On the other hand, if land purchase is necessary, it is often easier to acquire corner sites necessary for a roundabout than the long narrow strips needed when parallel widening of traffic signal approaches is carried out.

(b) Both conventional and mini-roundabouts have difficulty in dealing with unbalanced flows, especially during peak hours when the traffic entering on an arm is considerably greater than the traffic leaving by it. In such situations there is frequently a shortage of gaps in the circulating stream and, under off-side priority rule, delays may become excessive.

(c) Right-turning vehicles (left-hand rule of the road) cause difficulties with signal control when their numbers are large. Either late start or early cut-off facilities or a special phase must be provided causing reduced overall capacity at the junction. In such circumstances roundabouts offer advantages.

(d) Traffic signal control has difficulty in dealing with three-way junctions, especially where the flows are balanced. To a lesser extent this is true of junctions with five or more approaches.

From the point of view of traffic safety it has been found[2] that whenever a junction changed from one form of control to the other, a significant change in injury accident ate is observed; the accident rate on a roundabout is noted to be about two-thirds of the rate under signal control. This change is also true of pedestrian accidents and while pedestrians are often stated to feel less at risk at traffic signals their danger of injury at roundabouts is more apparent than real.

The traffic capacity of roundabouts

Prior to 1966 traffic entering and passing through a roundabout resolved the vehicle-to-vehicle interactions by requiring the traffic streams to weave one with another as they passed around the central island. In this situation the traffic design of round-abouts consisted of designing the individual weaving sections of the carriageway around the central island, an individual weaving section lying between adjacent entry arms.

These weaving sections were designed using a formula developed by Wardrop[3] which incorporated parameters that described the geometric shape of the weaving section and also the proportion of weaving vehicles in the section.

Without any positive control of traffic, roundabouts tended to 'lock' under heavy traffic conditions because vehicles already on the roundabout were prevented from leaving by vehicles attempting to enter. For these traffic conditions a solution was the use of long weaving sections and large roundabouts.

To overcome the problem of locking, the 'give way to traffic approaching from the right' rule was introduced. Subsequently a number of investigations showed that because traffic was no longer weaving around the central island the original formula was no longer valid and the correct approach was to consider the capacity of the roundabout as the capacity of the individual entries of the roundabout.

A particular difficulty in deriving a formula for the capacity of roundabouts is the variety of geometric designs found in practice. Designs range from conventional roundabouts with approximately parallel-sided rectangular-shaped weaving sections, to irregularly shaped central islands and entries on to the roundabout with widths rather similar to that of the approach carriageway, and to the offside priority roundabout with circular central islands and flared approaches.

Recommendations for determining the capacity of a roundabout entry are given by the Department of Transport[4] based on the research carried out by the Transport and Road Research Laboratory[5].

The predictive equation for entry capacity (Q_E) is given by $k(F - f_c Q_c)$ when $f_c Q_c$ is less than or equal to F (otherwise the entry capacity is zero) where

Q_E = the entry flow into the circulatory area in pcu/hour where a heavy goods vehicle equals two passenger cars),

Q_c = the flow in the circulating area in conflict with the entry in pcu/hour,

k = $1 - 0.00347(\Phi - 30) - 0.978[(1/r) - 0.05]$

F = $303X_2$

f_c = $0.210t_D(1 + 0.2X_2)$

t_D = $1 + 0.5/(1 + M)$

M = $\exp(D - 60)/10]$

X_2 = $v + (e - v)/(1 + 2S)$

S = $1.6(e - v)/l^l$

A description of these symbols is given in table 20.1 and a precise definition is given in figure 20.1.

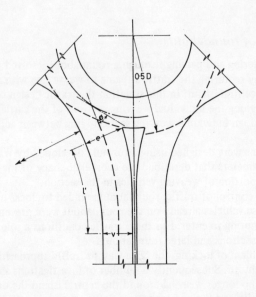

Figure 20.1

The equation for Q_E is applicable to all roundabout types except those that are incorporated into grade-separated intersections. In the latter case the term F is replaced by $1.11F$ and the term f_c is replaced by $1.4f_c$. The formula was derived from observations of traffic flow at roundabout intersections and the range of the geometric parameters at the observed roundabouts are given in table 20.1, together with the range of these parameters suggested for new design. It is also recommended that the circulatory carriageway width around the roundabout should be constant at 1 to 1.2 times the greatest entry width, subject to a maximum of 15 metres.

TABLE   20.1

Symbol	Description	Observed range	Recommended range for design
e	Entry width	3.6–16.5m	4.0–15.0m
v	Approach half-width	1.9–12.5m	2.0–7.3m
l^l	Average effective flare length	1–∞m	1.0–100.0m
S	Sharpness of flare	0–2.9m	
D	Inscribed circular diameter	13.5–171.6m	15–100m
Φ	Entry angle	0–77°	10–60°
r	Entry radius	3.4–∞m	6.0–100.0m

Whilst the use of the formula for Q_E will allow a check to be made that entry flows are below the entry capacity, there are difficulties when queueing commences. In these circumstances the entry and circulating flows are interdependent.

To enable delays to be predicted in this interactive situation the Transport and Road Research Laboratory have developed the computer program[6] ARCADY to model queues and delays at roundabouts.

Initially, there is assumed to be no circulating flow past the first entry, so that the entry flow will be either the entry demand or the entry capacity whichever is the least. This entry flow, after removal of vehicles that take the next exit, become the circulatory flow past the next entry. The entry flow at this next entry can then be determined by the circulating flow/entry flow relationship. This means that the circulatory flow and hence the entry flow at successive entries can be calculated in a clockwise manner. The process is iterated and the entry flows converge to their final values.

In the program successive short time intervals are considered during which the circulating flow, entry capacity and demand on the entering arm are known; hence the increase or decrease in queue length and the total queue length and delay can be calculated.

Input data for the program is of three types: reference information giving details of the computer run; date and names of roads at the intersection being studied; geometric characteristics of the junction and traffic flow information.

Output for the program includes tables of demand flows, capacities, queue lengths and delays for each time segment, together with a diagrammatic representation of the growth and decay of the queues on each arm.

Example

A roundabout at a grade-separated intersection has an inscribed circle diameter of 65 m, an entry width of 8.5 m and an approach half-width of 7.3 m. The effective length over which flare is developed is 30 m, the entry radius is 40 m and the entry angle is 60 degrees.

The design reference flows to be used for a preliminary design expressed as entry capacity (pcu per hour) are given in table 20.2. Calculated the reserve capacity of the intersection in these traffic conditions.

The flows given in table 20.2 are assigned to the roundabout as shown in figure 20.2. From this diagram the circulating flows around the central island are abstracted and these are shown in figure 20.3.

TABLE 20.2

From	To			
	N	S	E	W
N		850	200	100
S	700		450	250
E	150	350		700
W	350	450	350	

The capacity of an entry of a grade-separated roundabout may be calculated using the relationship

$$Q_e = k(F - F_c Q_c)$$

where

$$
\begin{aligned}
K &= 1 - 0.00347(\Phi - 30) - 0.978((^1/\text{r} - 0.05)) \\
  &= 1 - 0.00347(60 - 30) - 0.978((^1/40) - 0.05) \\
  &= 1 - 0.1041 + 0.02445 \\
  &= 0.92035. \\
S &= 1.6(e - v)/l^l \\
  &= 1.6(8.5 - 7.3)/30 \\
  &= 0.064. \\
X_2 &= v + (e - v)/(1 + 2s) \\
  &= 7.3 + (8.5 - 7.3)/(1 + 2 \times 0.064) \\
  &= 8.3638. \\
F &= 1.11(303 x_2) \\
  &= 1.11(303 \times 8.3628) \\
  &= 2813. \\
t_D &= 1 + 0.5/(1 + \exp((D - 60)/10)) \\
  &= 1 + 0.5/(1 + \exp((65 - 60)/10)) \\
  &= 1.1888. \\
f_c &= 1.4(0.210 t_D(1 + 0.2 x_2)) \\
  &= 1.4(0.210 \times 1.1888(1 + 0.2 \times 8.3638) \\
  &= 0.5846. \\
Q_e &= 0.92035 (2813 - 0.5846 Q_c) \\
  &= 2589 - 0.538 Q_c.
\end{aligned}
$$

Using this relationship between entry capacity and circulating flow the entry capacity for the N, S, E and W arms can be calculated using the values of circulating flow obtained from figure 20.3; calculated values are entered in table 20.3.

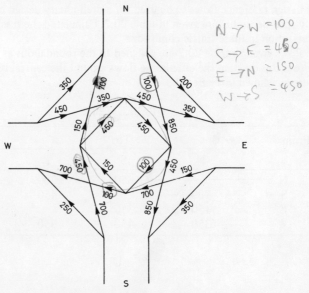

Figure 20.2

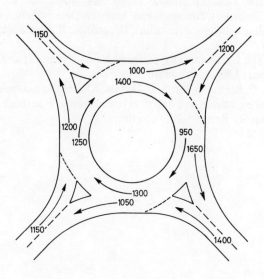

Figure 20.3

TABLE 20.3

Entry	Circulating flow (pcu/h)	Entry capacity (pcu/h)	Entry flow (pcu/h)	Reserve capacity (per cent)
N	1250	1917	1150	66.7
S	950	2078	1400	48.4
E	1400	1836	1200	53.0
W	1300	1890	1150	64.3

Comparison of the calculated entry capacities and the entry flows given in table 20.3 shows that the entry flow is lower than the calculated capacity for each of the entries. The reserve capacity is calculated from capacity minus flow divided by entry flow, expressed as a percentage.

References

1. R. S. Millard, Roundabouts and Signals, *Traff. Engng. Control* 13 (1971) 13–15
2. J. C. Tanner, Accidents before and after the provision and removal of automatic traffic signals, Department of Scientific and Industrial Research, RRL Note No. 2887 (1956)

3. J. G. Wardrop, The traffic capacity of weaving sections of roundabouts, Proceedings of the First International Conference on Operational Research, Oxford, 1957, English Universities Press, London, 1957, pp. 266–80
4. Department of Transport, Junctions and Accesses. Determination of size of roundabouts and major/minor junctions, Departmental Advice Note TA 23/81, London (1981)
5. R. M. Kimber, The traffic capacity of roundabouts, Transport and Road Research Laboratory Report LR942, Crowthorne (1980)
6. E. M. Hollis, M. C. Semmens and S. L. Dennis, ARCADY: a computer program to model capacities, queues and delays at roundabouts, Transport and Road Research Laboratory Report 940, Crowthorne (1980)

21

British and U.S. practice for determining the capacity of higher speed weaving sections

Weaving action takes place in many highway situations in addition to the conventional roundabout intersection where, it could be argued, priority control exists. It is therefore of interest to study the approach to the design of long weaving sections given in the Highway Capacity Manual[1]. The simplest long weaving section is a one-way highway at one end of which two highways merge and at the other end of which they diverge. Weaving is also inherent in many grade-separated interchanges in addition to the conventional flyover roundabout type of intersection. Where weaving is not found in the intersection then it frequently occurs on the highway between intersections when traffic enters the major route at one junction and leaves it at the next.

The Highway Capacity Manual defines several forms of weaving. Simple weaving is illustrated in figure 21.1. Figure 21.1(a) shows the basic weaving movement or simple weaving where all the vehicles entering the weaving section from either approach are destined to cross the path of all vehicles entering from the other approach, so that all the traffic weaves. Figure 21.1(b) illustrates dual purpose weaving in which both weaving and non-weaving traffic is combined. This is the form of weaving that is most prevalent and is of course the traffic action that takes place at a conventional roundabout. Where the volume of weaving traffic is high then more than two weaving lanes have to be provided and compound weaving takes place as shown in figure 21.1(c). By the provision of separate outer lanes for non-weaving traffic a higher level of service and an improvement in operating safety can be obtained.

Multiple weaving takes place when several weaving sections overlap each other as shown in figure 21.2. It occurs frequently at complex intersections in urban areas where a considerable number of traffic streams have to be collected and subsequently distributed.

The approach of the Highway Capacity Manual to the problem of the capacity of weaving sections is to consider separately the weaving traffic and the non-weaving traffic. At low volumes of flow these streams tend to separate from each other and even at high ratios of flow to capacity this general approach may be used.

189

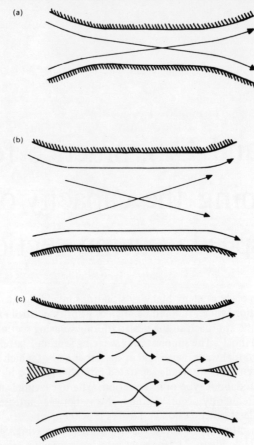

Figure 21.1 Some weaving sections (adapted from ref. 1)
(a) The basic weaving movement; (b) A dual purpose weaving section; (c) A compound weaving section

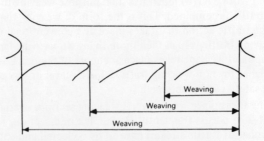

Figure 21.2 A compound weaving section

When the weaving movement is considered then it can be seen that all weaving vehicles have to cross a line running along the length of the weaving section from the nose of the entrance roadway to the nose of the exit roadway. The capacity of the weaving section cannot therefore exceed the capacity of a single lane. Should the flow of weaving vehicles exceed the capacity of a single lane then some vehicles are involved in two weaving movements. Where the weaving volume approaches a volume equal to twice the capacity of a traffic lane then a weaving length is required which is equal on this basis of reasoning to three times the length required for half this volume.

The effective length of a weaving section is also influenced, at least at the better levels of service, by the distance in advance of the weaving section at which drivers can see vehicles on the other approach road. This makes possible some speed adjustment so as to allow easier merging when a driver reaches the weaving section. The Highway Capacity Manual does not attempt to quantify these factors but the design charts used for predicting weaving capacity are based on adequate sight distances.

As has been previously stated, the approach of the Highway Capacity Manual is to divide flow through a weaving section into weaving and non-weaving flows. The capacity of the non-weaving sections can be calculated without difficulty from a knowledge of the capacity related to varying qualities of flow. The term quality of flow is used to describe traffic flow conditions through the weaving section and is related to the level of service on the approach highways.

Detailed studies were carried out throughout the United States on the relationship between the geometric features of weaving sections, traffic volumes and operating speeds. It was then found possible to represent these fundamental weaving relationships by means of a single basic weaving chart, which is reproduced from the Highway Capacity Manual as figure 21.3. This chart relates the weaving volumes possible at particular operating levels to the length of the section (the width of the section is

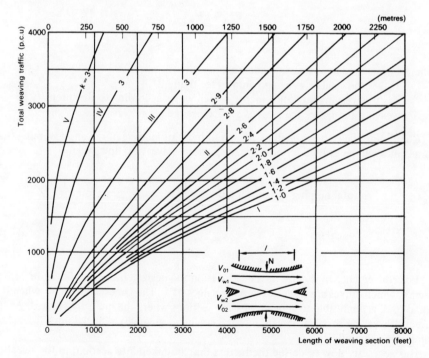

Figure 21.3 The capacity of weaving sections (adapted from ref. 1)

obtained from consideration of weaving and non-weaving volumes) and the lane service volume or capacity.

The number of lanes required for the outer or non-weaving flows may be calculated in the same manner as for any uninterrupted flow, by dividing the non-weaving volume by the appropriate lane service volume, that is

$$\frac{\text{flow}}{\text{lane service volume}} = \frac{V_{o1} + V_{o2}}{SV}$$

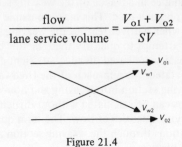

Figure 21.4

The number of lanes required for the weaving movements is similarly calculated using the same lane service volume but including a correction factor which takes into account the fact that more width is required for weaving than for non-weaving flow. Using data obtained from traffic studies, the number of lanes required for weaving traffic may be calculated as

$$\frac{V_{w1} + kV_{w2}}{SV}$$

where V_{w1} is the larger weaving volume,
V_{w2} is the smaller weaving volume (see figure 21.4)
k is the weaving influence factor, which varies according to the intensity of weaving and hence the level of service desired.

Combining the two expressions for the number of lanes required and assuming that some lanes will be utilised by both weaving and non-weaving flows then

$$\text{total number of lanes required} = \frac{V_{w1} + kV_{w2} + V_{o1} + V_{o2}}{SV}$$

$$= \frac{V + (k-1)V_{w2}}{SV}$$

For convenience in the use of the expression the design chart given in figure 21.3 is used in which curves are given for varying values of k. The largest value of k, when it is equal to 3, is applicable to shorter weaving sections where the intensity is great. As the effect of weaving becomes less intense the k-value is gradually reduced until it reaches a minimum value of 1.

It is necessary now to define the factors that are taken into account in the selection of the appropriate intensity of weaving for a wide range of highway types. As each type of highway has its own levels of service criteria it was not found possible to apply the basic levels of service to the weaving chart. Instead the term levels of quality of flow is used, and by the use of table 21.1 the appropriate quality of flow can be obtained if the highway type and the level of service on the highway are known.

TABLE 21.1   Levels of service and qualities of flow

| Level of service | Multilane highways | | Two-lane rural highways | Urban and suburban arterials |
	Highway proper	Interchange		
A	I-II	II-III	II	III-IV
B	II	III	II-III	III-IV
C	II-III	III-IV	III	IV
D	III-IV	IV	IV	IV
E	IV-V	V	V	V
F	←	Not satisfactory		→

* Relationships below line are not considered suitable for design purposes.

A description of the traffic flow conditions for the five qualities of flow is given below.

I   Operating conditions and speeds approach those found under free flow conditions without weaving. The effect of weaving is slight; speeds of 50 m.p.h. or more are possible.

II   Operating conditions and speeds only slightly restricted. Some speed variations occur; speeds of 45–50 m.p.h. possible.

III   Some drivers affected by other vehicles; flow conditions not unreasonable for situations where speeds on the approach highways are 50 m.p.h.; speed variations occur.

IV   Occasional slow downs and some restrictions on manoeuvrability. Suitable where approach highway speeds are 40 m.p.h.

V   Capacity conditions, speeds 30–20 m.p.h.; slow operation.

While it is physically difficult to maintain the same operating speed through weaving sections as on the approach highways, the weaving sections should provide at least the same capacity as the approach highways. Nevertheless every attempt should be made to obtain minimum loss of quality of flow through the weaving sections but as long as the quantity of flow is not much less than the level of service on the approach highways and it does not occur at too frequent intervals, then it may prove acceptable.

Definition of length of weaving sections

The lengths of weaving sections as given in figure 21.3 are defined as shown in figure 21.5, and in the analysis of existing weaving sections or the design of new sections the length must be taken as shown.

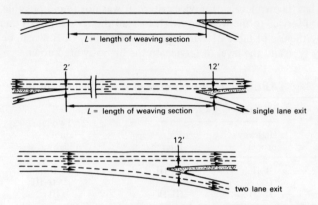

Figure 21.5 The measurement of weaving section lengths (adapted from ref. 1)

Service volumes

The service volumes (*SV*) used to calculate the number of lanes required is normally the average service volume per lane for the level of service chosen for the approach and exit roadways, taking into account in the case of levels of service C and D the appropriate peak hour factors. Corrections have to be made to these lane service volumes for traffic composition, lane width, gradient and other traffic and roadway factors.

The Highway Capacity Manual has established certain maximum values for the service volumes to be used in conjunction with the weaving qualities of flow. These values, which are given in table 21.2, still represent ideal roadway and traffic conditions.

TABLE 21.2   Relationship between quality
of flow and maximum lane service volume in
weaving sections

Quality of flow	Maximum value of lane service volume passenger cars/hour
I	2000
II	1900
III	1800
IV	1700
V	1600

The use of the weaving chart (figure 21.3) can be illustrated by the following example on the calculation of weaving section dimensions.

What are the minimum length and width requirements for an intersection in a rural area with traffic flows as given in figure 21.6? The approach roadways have a level of service B, there is negligible commercial traffic and the maximum service volume for this level of service is 2000 p.c.u./hour for a two-lane carriageway.

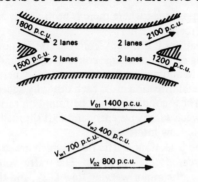

Figure 21.6 A multiple weaving situation

From table 21.1 when the level of service is B on a rural highway, the quality of flow may be taken as II. Taking $k = 2.6$ in figure 21.3 for a total weaving volume of 1100 equivalent private cars the required length of the weaving section is 1500 ft.

$$\text{width} = \frac{V + (k-1)V_{w2}}{SV} = \frac{3300 + (2.6-1)400}{1000}$$

$$= 3.9, \text{ that is 4 lanes.}$$

Sections out of realm of weaving

Weaving intensity increases as weaving length decreases and conversely as weaving length increases weaving intensity decreases. For this reason there must come a point where weaving is spread out over such a length of highway that the traffic action is similar to normal lane changing.

TABLE 21.3

Volume weaving p.c.u./hour	Min. length of section (ft)
500	1000
1000	2400
1500	4000
2000	6000

While the precise point at which weaving may be disregarded is not known, table 21.3 gives an indication of the limits beyond which weaving need be considered. These figures form curve I on the weaving chart and flow conditions to the right of curve I may be considered to be beyond the realm of weaving.

Design of weaving areas in Great Britain

The design of weaving areas for motorways and all purpose roads in Great Britain is described in the Department of Transport Technical Memorandum H12/76.[2] The method, which is based on the Highway Capacity Manual approach previously described, is applicable to both rural and urban situations where the weaving area is on a motorway or a high quality all-purpose road where there is no frontage access, no standing vehicles and negligible crossing traffic.

Weaving in these situations is different from the traffic movements which might be expected on the weaving section of a large conventional roundabout. Traffic movements applicable to Technical Memorandum H12/76 are those where two traffic streams each moving in the same general direction cross each other by successive merging and diverging manoeuvres. These conditions may be found on carriageways between adjacent junctions, between successive entry and exit sliproads at junctions, and on the link roads within free flow interchanges.

Design is based on the peak hour traffic flows which are expected in the design year, normally the fifteenth year after construction. The traffic volumes are given as the major weaving flow Q_{w1}, the minor weaving flow Q_{w2}, and the non-weaving flow Q_{nw}. The flow is expressed in vehicles with the percentage of heavy vehicles (greater than 1.5 tonne unladen) given. This is converted to a standard composition of traffic which is taken as 15 per cent of heavy vehicles together with a weaving length gradient of up to 1 per cent uphill. The average gradient is normally measured over the length of the weaving section plus a 0.5 km length of mainline highway upstream of the merge nose. Corrections to convert traffic flows to standard composition range from a reduction of 8 per cent when there is 5 per cent of heavy vehicles and a downhill, level, or up to 1 per cent uphill gradient to an increase of 40 per cent when there is 40 per cent of heavy vehicles and an uphill gradient of 3 per cent.

It is also necessary to know the design speed V and design flow D for the mainline upstream of the weaving area. The design flow is the allowable peak hourly design flow per lane based on tables 11.2 and 11.3.

The minimum length of weaving section is obtained by the use of figure 21.7 which relates the total weaving flow, that is the sum of the major and minor weaving movements to the minimum length of the weaving section for varying values of the ratio of design flow to design speed. An absolute minimum length of weaving section which is related

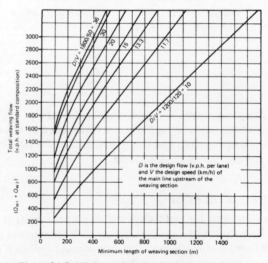

Figure 21.7 Minimum length of weaving section

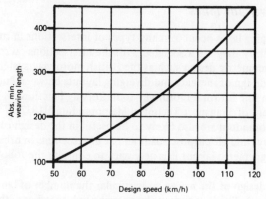

Figure 21.8

to design speed for the main line flow upstream of the weaving section is also specified as is obtained for rural highways from figure 21.8. The greater of these two lengths is selected as the minimum acceptable length of the weaving section L_{min}. In an urban area the minimum weaving length is recommended as 180 m.

To calculate the width of the weaving area, a weighting factor is applied to the minor weaving flow to represent the reduction in capacity caused by the weaving operation. This factor has a maximum value of 3 when the actual length L_{act} of the weaving section is equal to the minimum value L_{min} and reduces to a limiting value of 1 as the actual length of the weaving section increases according to the relationship

$$\text{weighting factor} = \left(\frac{2 \times L_{min}}{L_{act}} + 1\right)$$

The number of lanes N required in the weaving area can now be obtained from the expression

$$N = \frac{Q_{nw} + Q_{w1}}{D} + \left(\frac{2 \times L_{min}}{L_{act}} + 1\right)\frac{Q_{w2}}{D}$$

The Technical Memorandum gives advice on the decision which has usually to be made whether to round up or down the value of N which is obtained by calculation. The rounding down of the value may, in some circumstances, result in a long weaving area length; in other circumstances the decision is obvious, such as when there is a high fractional part of N combined with high weaving volumes and design flows. If the length of the weaving area is fixed by the position of slip roads, then the following factors should also be considered: whether the traffic is likely to contain a high percentage of drivers who are unfamiliar with the junction; the cost and availability of land; the standard of the geometric design; and the actual number of lanes which will finally result from the rounding process.

Merging and diverging lanes

Merging and diverging lanes occur in many types of intersections in rural and urban areas. Most frequently they are associated with the intersections on routes with a high standard of geometric design such as the British motorway system.

Advice on the design of merging and diverging layouts between entry and exit links and the main line for rural motorway to motorway junctions in Great Britain is given in Department of Transport Technical Memorandum H18/75.[3] The design princip given in this memorandum are also likely to be useful in the design of intersections between motorways and all purpose roads and at intersections in urban areas.

When designing a merge layout it is necessary to consider the following factors:

(i) the geometric design of the merge, in particular the number of lanes on the upstream and downstream main line link and on the entering link together with the vertical and horizontal alignment through the merge area;
(ii) the flows and composition of the traffic on the main line and merging link for the design year using the worst combination of peak hour flows;
(iii) the Peak to Daily Flow Ratio (PDR) for the main line prior to the merge and for th merge flow if it is heavy (see p. 98).

When designing a diverge layout similar factors must be considered as given for the merge layout with the substitution of diverge link for merge link. As in the design of weaving areas, the design year traffic flows are corrected to a standard composition of 1 per cent heavy vehicles and a gradient of downhill, level or up to 1 per cent uphill.

The design procedure is based on maximum working levels of flow of 1600 vehicles per hour per lane, for multi lane highways and 1200 vehicles per hour per lane, for single lane carriageways. Detailed recommendations are given in the memorandum for situations where the standard peak hourly flow level is exceeded.

The type of merge or diverge layout suitable for various upstream main line flows and entry flows or downstream main line flows and exit flows is indicated by figures 21.9 and 21.10 respectively. These diagrams indicate flow regions which are defined in Table 21.4.

These merging and diverging diagrams were obtained by considering the following limiting flow conditions. For the merging diagram

(i) the capacity of the main carriageway upstream of the merge at 1600 vehicles per hour, per lane;
(ii) the capacity of the entry link, 3200 vehicles per hour for a two lane link, and 1200 vehicles per hour for a single lane link;
(iii) the capacity of the main carriageway downstream of the entry link;
(iv) the principle that the entry flow should not exceed the upstream main carriageway
(v) the limitation due to the merging process itself, the relationship between the nearsid main carriageway flow, the total main carriageway flow and the merging flow is based on observations contained in the Highway Capacity Manual.[1]

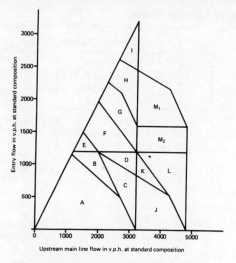

Figure 21.9 Merging diagram (from ref. 3)

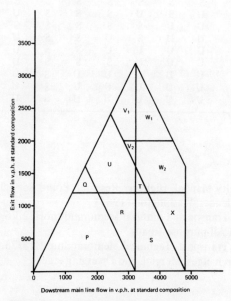

Figure 21. 10 Diverging diagram (from ref. 3)

TABLE 21.4 Merge and diverge designs

		Number of lanes										
Upstream main line		2	2	2	2	3	3	3	3	2	2	3
Link		1	1	2	2	1	1	2	2	1	2	1
Entry		1	2	2	2	1	2	2	2	1	2	1
Downstream main line		2	2	2	3	3	3	3	4	3	4	4
	A	S	S	S	S	S	S	S	S	S	S	S
	B	U	S	S	S	S	S	S	S	S	S	S
	C	U	U	U	S	S	S	S	S	S	S	S
	D	U	U	U	S	U	S	S	S	S	S	S
	E	U	U	S	S	U	U	S	S	U	S	U
Flow	F	U	U	U	S	U	U	S	S	U	S	U
region	G	U	U	U	S	U	U	U	S	U	S	U
	H	U	U	U	U	U	U	U	S	U	S	U
	I	U	U	U	U	U	U	U	U	U	S	U
	J	U	U	U	U	S	S	S	S	U	U	S
	K	U	U	U	U	U	S	S	S	U	U	S
	L	U	U	U	U	U	U	U	S	U	U	S
	M	U	U	U	U	U	U	U	S	U	U	U
Upstream main line		2	2	3	3	3	4	3	4	4		
Link		1	2	1	2	2	2	1	1	2		
Downstream main line		2	2	3	2	3	3	2	3	2		
	P	S	S	S	S	S	S	S	S	S		
	Q	U	S	U	S	S	S	U	U	S		
	R	U	U	S	S	S	S	S	S	S		
	S	U	U	S	U	S	S	U	S	U		
	T	U	U	U	U	S	S	U	U	U		
	U	U	U	U	S	S	S	U	U	S		
	V	U	U	U	U	U	S	U	U	S		
	W	U	U	U	U	U	S	U	U	U		
	X	U	U	U	U	U	S	U	S	U		

S = Satisfactory
U = Unsatisfactory

References

1. Highway Capacity Manual. Highway Research Board Special Report 87 National Academy of Sciences, Washington (1965)
2. Department of Transport. Technical Memorandum H12/76, Weaving Areas for Motorways and All-purpose Roads
3. Department of Transport. Technical Memorandum H18/75, Rural Motorway to Motorway Interchanges, Merging and Diverging lanes

22

Queueing processes in traffic flow

Queueing theory originally developed by A. K. Erlang in 1909 has found widespread application in the problems of highway traffic flow[1].

In any highway traffic situation it is necessary to know: the distribution of vehicle arrivals into the queueing system; whether the source of vehicle arrivals is finite or infinite; whether queue discipline is first come first served, priority or random selection; the number of service stations whereby the vehicle may exit from the system and the distribution of service times for each service station.

A typical system occurs at the entry or exit of a stream of vehicles into or from a parking garage. The vehicles arrive at the parking garage at random, form a queue and enter or leave the garage on a first-come first-served basis. The times required by vehicles to pass through a garage entrance or exit form an approximation to an exponential distribution.

While the input distribution is assumed as random and the service time distribution is exponential in the above example, the application of queueing theory to traffic engineering has been mainly developed around the regular, random and Erlang distributions.

When vehicles arrived at random the numbers of vehicles arriving in successive intervals of time can be represented by the Poisson distribution and an understanding of queueing theory can be obtained from this simple case of Poisson-distributed vehicle arrivals in a single lane, in which vehicles depart with an exponentially distributed service rate.

Consider a traffic queue where $P(n, t + dt)$ is the probability that the queue contains n vehicles $(n > 0)$ at time $t + dt$. There are three ways in which the system could have reached this state if it is assumed that dt is so small that only one vehicle could have arrived or departed. These are:

(a) a vehicle did not arrive or depart in time t to $t + dt$;
(b) the queue contained $n - 1$ vehicles at time t and one arrived in time dt;
(c) the queue contained $n + 1$ vehicles at time t and one departed in time dt.

Now with Poisson distributed arrivals

$$P(n) = \frac{(qt)^n}{n!} \exp(-qt) \tag{22.1}$$

201

where $P(n)$ is the probability of n vehicles arriving in time t when the mean rate of vehicle arrival is q. From equation 22.1

$$P(0) = \exp(-q\ dt) \qquad (22.2)$$

where $P(0)$ is the probability of zero arrivals in time dt and

$$P(1) = (q\ dt)\exp(-q\ dt) \qquad (22.3)$$

where $P(1)$ is the probability of one arrival in time dt.

Expanding equation 22.2

$$P(0) = (1 - q\ dt + q^2\ dt^2/2!\ \ldots)$$

Expanding equation 22.3

$$P(1) = q\ dt(1 - q\ dt + q^2\ dt^2/2!\ \ldots)$$

Neglecting second and higher powers of dt

$$P(0) = (1 - q\ dt)$$

and

$$P(1) = q\ dt$$

Similarly the probability of 0 and 1 departures from the queue are

$$P(0) = (1 - Q\ dt)$$

and

$$P(1) = Q\ dt$$

where Q is the mean rate of departure from the queue. Where $n > 0$ the system can reach a state of n vehicles at time $t + dt$ in the following manner

$$P(n, t + dt) = P(n, t) \times P\ (\text{a vehicle does not arrive or depart})$$
$$+ P(n - 1, t) \times P\ (\text{a vehicle arrives})$$
$$+ P(n + 1, t) \times P\ (\text{a vehicle departs})$$
$$= P(n, t)((1 - q\ dt)(1 - Q\ dt))$$
$$+ P(n - 1, t)(q\ dt) + P(n + 1, t)(Q\ dt)$$

Ignoring second and higher powers of dt

$$P(n, t + dt) = P(n, t)(1 - q\ dt - Q\ dt) + P(n - 1, t)(q\ dt) + P(n + 1, t)(Q\ dt)$$
$$\frac{P(n, t + dt) - P(n, t)}{dt} = -P(n, t)(q + Q) + P(n - 1, t)(q) + P(n + 1, t)(Q)$$
$$(22.4)$$

In the limit for the steady-state solution the rate of change is zero. Hence

$$0 = -P(n)(q + Q) + P(n - 1)(q) + P(n + 1)(Q)$$

or

$$(1 + q/Q)Pn = (q/Q)P(n - 1) + P(n + 1) \qquad (22.5)$$

Similarly when $n = 0$, there are two ways in which the queue can contain n vehicles at time $t + dt$, a change of type (a) or a change of type (c)

$$P(0, t + dt) = P(0, t)(1 - q\, dt) + P(0 + 1, t)(Q\, dt)$$

$$\therefore \frac{P(0, t + dt) - P(0, t)}{dt} = P(0 + 1, t)(Qt) - P(0, t)(qt)$$

As before the steady state of the queue probability of n vehicles in the system is

$$P(1) = P(0)(q/Q)$$

From equation 22.5 when $n = 1$

$$P(2) = (q/Q)^2 P(0) \tag{22.6}$$

Similarly when $n = 2$

$$P(3) = (q/Q)^3 P(0)$$

By induction it can be shown that

$$P(n) = (q/Q)^n P(0) \tag{22.7}$$

When the queue size may be infinite

$$P(0) + P(1) + P(2) + P(3) + \cdots + P(\infty) = 1$$

From equation 22.7

$$P(0) + (q/Q)P(0) + (q/Q)^2 P(0) + (q/Q)^3 P(0) + \cdots + (q/Q)^\infty P(0) = 1$$

$$P(0)(1/(1 - q/Q)) = 1$$

$$P(0) = (1 - q/Q)$$

Also

$$P(n) = (q/Q)^n (1 - q/Q) \tag{22.8}$$

The expected number in the queue $E(n)$ is given by

$$E(n) = \sum_{n=0}^{\infty} nP(n)$$

$$= 0 \times P(0) + 1 \times P(1) + 2 \times P(2) + \cdots + nP(n)$$

$$= (q/Q)P(0) + 2(q/Q)^2 P(0) + \cdots + n(q/Q)^n P(0)$$

$$= (q/Q)P(0)(1 + 2(q/Q) + \cdots + n(q/Q)^{n-1})$$

$$= (q/Q)P(0)(1/(1 - q/Q)^2)$$

$$= q/(Q - q) \tag{22.9}$$

Because there is a probability that the queue will be zero the mean queue length $E(m)$ will not exactly be one less than the mean number in the queue $E(n)$

$$E(m) = \sum_{n=1}^{\infty} (n - 1)P(n)$$

$$= \sum_{n=0}^{\infty} P(n)n - P(n) + P(0)$$

$$= E(n) - q/Q \tag{22.10}$$

As well as the expected number in the queue and the mean queue length, the waiting time w before being taken into service and the total time in the queue v are of considerable importance in the study of traffic phenomena.

The waiting time distribution may be considered in two parts.

Firstly there is the probability that the waiting time will be zero, which means that a queue does not exist, and

$$P(0) = 1 - q/Q, \qquad n = 0$$

Secondly there is the probability that the waiting time for a vehicle is between time w and time $w + dw$

$$P(w < \text{wait} < w + dw) = f(w)\, dw, \qquad n > 0 \tag{22.11}$$

Such a delay is possible as long as there is a vehicle in service, which may be expressed as

$$P(n \geqslant 1) = \sum_{n=1}^{\infty} Pn \tag{22.12}$$

For the waiting time for a vehicle to be exactly between w and $w + dw$ all the vehicles in the queue ahead of the one being considered except the one immediately ahead, must depart in time w and the one immediately ahead must be served in time dw. This is the product of the two probabilities

$$P(n - 1, w) = \frac{(Qw)^{n-1}}{(n - 1)!} \exp(-Qw)$$

from equation 22.1 and

$$P(1, dw) = Q\, dw$$

Substituting these probabilities in equation 22.11 and summing over equation 22.12

$$f(w)\, dw = \sum_{n=1}^{\infty} P(n)P(n - 1, w)P(1, dw)$$

$$= \sum_{n=1}^{\infty} (q/Q)^n (1 - q/Q) \times \frac{Q\, dw}{(n - 1)!} (Qw)^{n-1} \exp(-Qw)$$

$$= q(1 - q/Q)\, dw \exp(-Qw) \sum_{n=1}^{\infty} \frac{(wq)^{n-1}}{(n - 1)!}$$

$$f(w) = (q/Q)(Q - q) \exp(-w(Q - q)) \qquad w > 0 \tag{22.13}$$

The moment generating function for waiting times is given by

$$M_w(\theta) = \int_0^\infty \exp(\theta w) f(w) \, dw$$

$$= \int_0^\infty \exp(\theta w)(q/Q)(Q-q) \exp(-w(Q-q)) \, dw$$

$$= (q/Q)(Q-q) \int_0^\infty \exp(-w(Q-q-\theta)) \, dw$$

$$= (q/Q)(Q-q)/(Q-q-\theta)$$

$$M'w(0) = E(w) = q/[Q(Q-q)] \tag{22.14}$$

The average time an arrival spends in the queue is given by $E(w)$ plus the average service time $1/Q$

$$E(v) = 1/(Q-q) \tag{22.15}$$

For more generalised cases when the service time can no longer be described by a negative exponential distribution the expected number in the queue when arrivals are random is given by

$$E(n) = \frac{q}{Q} + \left(\frac{q}{Q}\right)^2 (1 + C^2) \,/\, \left[2\left(1 - \frac{q}{Q}\right)\right] \tag{22.16}$$

where C is the coefficient of variation of the service time distribution, that is the ratio of the standard deviation to the mean.

If the service is exponential then $C^2 = 1$ and equation 22.16 reduces to equation 22.9. If the service is regular $C^2 = 0$ and

$$E(n) = \frac{q}{Q}\left(1 - \frac{q}{2Q}\right) \,/\, \left(1 - \frac{q}{Q}\right) \tag{22.17}$$

In this case it has been shown that the average time a vehicle spends queueing is given by

$$E(w) = q/2Q(Q-q) \tag{22.18}$$

To illustrate the use of queueing theory in highway traffic flow it is necessary to find some situation in which vehicles are delayed and allowed to proceed in accordance with the simple situations previously considered.

Typical situations occur when vehicles have to stop to enter or leave parking facilities during a period of exceptional demand. Queues can frequently be observed at the entrance to or exit from car parks on public holidays, at weekends at the coast and at sporting events.

It is desirable that the rate of arrival q is reasonably constant and, to make this possible, observations may be divided into shorter periods of time, each with separate value of q.

In some cases the service rate may vary with demand but where queues are forming it is usually possible to assume that the service rate is approximately constant.

As an example of the application of queueing theory observations were made of the number of vehicles waiting to enter two parking areas. One was a multistorey parking garage equipped with automatic entry control equipment and the other was a surface car park where drivers paid the attendant as they entered. In both instances vehicles were delayed as they queued to enter the park and drivers on the approach had considered only one choice of entry gate.

Observations made of vehicle arrivals at the entrance to the surface car park are included to demonstrate the statistical technique whereby vehicle arrivals are shown to be random or non-random.

Where vehicles arrive at random then the numbers of vehicles arriving in successive time intervals may be represented by the Poisson distribution.

Then the probability of n vehicles arriving in a given interval of time t may be calculated from

$$P(n) = \frac{(qt)^n \exp{(-qt)}}{n!} \tag{22.19}$$

This distribution is often referred to as the counting distribution because it describes the number of vehicles arriving at a given point on the highway.

The numbers of vehicles arriving at the entrance to the surface car park or at the end of the queue in successive 60-second intervals were observed. At this situation marked changes in the arrival rate were not expected and so observations were continued for a period of 3000 seconds and the mean arrival rate taken as q.

To test the form of the arrival distribution and also to determine the mean rate of arrival the observations were tabulated as shown in table 22.1.

TABLE 22.1   Vehicle arrival distribution at the entrance to a surface car park

No. of vehicles arriving in a 60 s interval	Frequency of observed intervals		Theoretical frequency $(P(n) \Sigma f_0)$	Chi-squared
(n)	(f_0)	$(f_0 n)$	(f_t)	(x^2)
0	1 ⎫	0	0 ⎫	
1	2 ⎬	2	2·40 ⎬	0·01
2	4 ⎭	8	4·86 ⎭	
3	9	27	7·71	0·22
4	10	40	9·18	0·07
5	7	35	8·74	0·35
6	6	36	6·94	0·13
7	4 ⎫	28	4·72 ⎫	
8	3 ⎪	24	3·83 ⎪	
9	2 ⎬	18	1·49 ⎬	0·01
10	1 ⎪	10	0·71 ⎪	
11	1 ⎭	11	0 ⎭	
Σ 50		Σ 239		Σ 0·79

$$\text{mean headway} = \frac{60 \times 50}{239} = 12 \cdot 6 \text{ s}$$

$$q = 1/12 \cdot 6 = 0 \cdot 08 \text{ veh/s}$$

$$\text{arrival volume} = 3600/12 \cdot 6 = 286 \text{ veh/h}$$

Two further sets of data were obtained at this site and the values of both q and Q derived from the observations are included in table 22.4.

It can be seen from tables of chi-squared that there is no significant difference at the 95 per cent level between the observed and the theoretical distributions.

The mean time taken for a vehicle to enter the car park, the service time, was observed as the time interval between successive vehicles in the queue moving away from the attendant or passing beneath the raised barrier in the case of the parking garage.

Table 22.2 gives the observed distribution of service times at the entrance to the surface car park. The general exponential nature of the service time can be seen. Over 50 per cent of drivers are able to pay the attendant and receive a receipt in less than 8 seconds. A smaller number of drivers require a longer period to tender the fee and receive change. No driver takes longer than 28 seconds to enter the car park after arriving at the entrance.

Where the service time distibution is exponential the probability of drivers requiring service times between the class limits $t - \Delta t/2$ and $t + \Delta t/2$ may be calculated from

$$\frac{\Delta t}{t_s} \exp - \frac{t}{t_s}$$

where Δt is the class interval,

t is the class mark,

$\bar{t}_s$ is the mean service time.

Observed and theoretical values are compared in table 22.2 and it can be seen from tables of χ^2 that there is no significant difference at the 5 per cent level.

TABLE 22.2   Vehicle service distribution at entrance to surface car park

Service time class interval (seconds)	Observed frequency (f_0)	$f_0 t$	Theoretical frequency f_t	χ^2
0–3·9	80	160	87	0·56
4–7·9	53	318	55	0·07
8–11·9	41	410	35	1·03
12–15·9	28	392	22	1·64
16–19·9	18	324	14	0·29
20–23·9	12	264	9	1·00
24–27·9	8 ⎱	208	6 ⎱	4·00
28	0 ⎰	0	10 ⎰	
	$\Sigma\, 240$	$\Sigma\, 2076$		$\Sigma\, 8·59$

$\bar{t}_s = 2076/240 = 8·7$ seconds

Observations of the service time at the entrance to the multistorey parking garage show a considerably different form of distribution. In this case the drivers had only to drive into the entrance bay, take a ticket and drive into the garage when the barrier had been automatically raised. Table 22.3 gives details of these observed service times for one of the 3000-second periods of observation made at this garage.

TABLE 22.3   Vehicle service distribution at entrance to multistorey car park

Service time class interval (seconds)	Observed frequency f_0	$f_0 t$
0–3·9	27	54
4–7·9	120	720
8–11·9	31	310
12–15·9	5	70
	$\Sigma\,183$	$\Sigma\,1154$

$$\bar{t}_s = 1154/183 = 6\cdot3 \text{ seconds}$$

At the same time as observations were made of the arrival and service time distributions a note was made of the queue length at 100-second intervals.

The average delay to vehicles entering the car park was calculated from

$$\frac{\Sigma \text{ sum of queue lengths} \times 100}{\text{no. of vehicles arriving}}$$

The observed delay to queueing vehicles for each 3000-second period of observation together with the mean arrival and service rates are given in table 22.4.

Theoretical delays were calculated assuming an exponential distribution of service times (equation 22.14) and a regular distribution of service times (equation 22.18).

TABLE 22.4

Arrival volume (veh/h)	q (veh/s)	Mean service time	Q (veh/s)	Observed average delay	Theoretical delay Eq. 22.14	Theoretical delay Eq. 22.18
Surface car park						
288	0·0800	8·7	0·1149	25	20	10
293	0·0814	8·7	0·1149	23	21	11
327	0·0908	8·7	0·1149	35	32	16
Multistorey car park						
506	0·1406	6·3	0·1587	41	55	24
497	0·1381	6·3	0·1587	33	49	21
533	0·1481	6·3	0·1587	61	94	44

It can be seen from table 22.4 that observed delays at the entrance to the surface car park can be approximately represented by equation 22.14. On the other hand delays at the entrance to the multistorey car park, where service is approximately regular because it is only necessary to take a ticket and move through a barrier, exhibit characteristics midway between those given by equations 22.14 and 22.18.

Using the equations derived from queueing theory it would be possible to form an estimate of delays of higher arrival volumes. This allows a balance to be obtained between delays to queueing vehicles and the cost of opening additional entrances.

Suggested reading

1. P. M. Morse. *Queues, Inventories and Maintenance*, Wiley, New York (1963)
2. D. R. Cox and W. L. Smith, *Queues*, Methuen, London (1961)
3. W. D. Ashton. *The Theory of Road Traffic Flow*, Methuen, London (1966)

Problems

1. A census point is set up on a highway where vehicle arrivals may be assumed to be random and the one-way traffic volume is 720 veh/h. All vehicles are required to stop at the census point while a tag is attached, the operation taking a uniform time interval of 4 s.

(a) Is the expected number of vehicles waiting at the census point 1·4, 2·4 or 3·0?

(b) Is the average waiting time of a vehicle at the census point 6·5 s, 8·0 s or 9·4 s?

(c) The enumerators are replaced by untrained staff and the time taken to attach a tag now has the same mean value but the standard deviation of the time taken is noted to be 2 s. Will the number of vehicles waiting at the checkpoint be increased by more than 50 per cent?

Solutions

1. (a) At a queueing situation in which arrivals are randomly distributed and where the service times are uniform then the expected number of vehicles waiting at the census point is given by equation 22.17

$$E(n) = \frac{q}{Q}\left(1 - \frac{q}{2Q}\right)\bigg/\left(1 - \frac{q}{Q}\right)$$

where q is the mean rate of arrival of vehicles,
  Q is the mean rate of departure of vehicles.
  In this example

$$q = 720/3600 \text{ veh/s}$$

$$= 0·2 \text{ veh/s}$$

$$Q = 1/4 \text{ veh/s}$$

$$= 0·25 \text{ veh/s}$$

$$E(n) = \frac{0·2}{0·25}\left(1 - \frac{0·2}{2 \times 0·25}\right)\bigg/\left(1 - \frac{0·2}{0·25}\right)$$

$$= 0·8(1 - 0·4)/(1 - 0·8)$$

$$= 2·4$$

(b) For randomly distributed arrivals and uniform service times the average waiting time is given by equation 22.18

$$E(w) = q/2Q(Q - q)$$

$$= \frac{0 \cdot 2}{2 \times 0 \cdot 25(0 \cdot 25 - 0 \cdot 2)}$$

$$= 8 \text{ s}$$

(c) In the general case in which arrivals are random and service times are distributed with the mean service time and the standard deviation of service times known, then the expected number of vehicles waiting to be serviced is given by equation 22.16

$$E(n) = \frac{q}{Q} + \left(\frac{q}{Q}\right)^2 (1 + C^2)/\left[2\left(1 - \frac{q}{Q}\right)\right]$$

where C is the coefficient of variation of the service time distribution.

$$E(n) = \frac{0 \cdot 2}{0 \cdot 25} + \left(\frac{0 \cdot 2}{0 \cdot 25}\right)^2 (1 + (\tfrac{2}{4})^2)/\left[2\left(1 - \frac{0 \cdot 2}{0 \cdot 25}\right)\right]$$

$$= 2 \cdot 7 \text{ vehicles}$$

As the previous number of vehicles waiting at the checkpoint was $2 \cdot 4$, the change in the enumerating staff has not resulted in an increase of more than 50 per cent in the number of waiting vehicles.

23

New forms of single-level intersections

As the number of vehicles in use increases in all developing countries the most severe congestion is being experienced in urban areas where the number of junctions severely restricts traffic flow. While restraint on car use can well be expected in such areas in the future, it is still necessary to increase the capacity of junctions for a balanced highway system.

The usual solution to the problem of junction capacity has been either to make the area of the junction larger, or to adopt grade separation. The former is difficult to achieve in congested urban areas and the latter is extremely expensive. It is therefore important to investigate every possible means of increasing the capacity of single-level junctions so that delays are reduced.

An investigation into ways of obtaining the highest possible capacity from a given area of junction has been carried out at the Transport and Road Research Laboratory[1]. In contrast to many previous experiments, which had as their object improvements in the capacity of a fixed shape of intersection, the shape of the intersection in these experiments was varied but its plan area was kept constant.

The experiments were carried out on the Transport and Road Research Laboratory test track using up to 150 vehicles and drivers.

A variety of junction shapes was tested varying from a straight crossroads with square corners and with no approach widening to a roughly square junction with the approach roads on the diagonals. Differing proportions of turning movements were used and the traffic flow was composed of differing vehicle types.

Both signal and priority control were used. With signal control both hooking and non-hooking right turns were used, while for priority control the size of the central island was varied from a reasonable maximum down to virtually no island at all. In some circumstances vehicles were allowed to pass the central island point on the wrong side while both nearside and offside priority control were used.

A wide variety of internal layout and shape of widening was used, modifying a straight crossroads with rounded corners by the addition of approximately 1500 m^2 of road area in several ways. The effect of both increasing the size of an intersection and the shape of the widening is illustrated in figure 23.1. A basic crossroads with 31·8 m^2 of area added by the use of 6·1 m corner radii was found to have a maximum

211

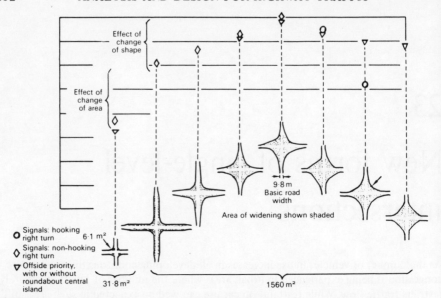

Figure 23.1 The effect of area and shape of widening for the same basic road width (adapted from ref. 1); ○ signals: hooking right turn; ◇ signals: non-hooking right turn; ▽ offside priority, with or without roundabout central island

observed capacity of between 3000 and 4000 p.c.u./h with both signal and offside priority control. The effect of internal layout is clearly illustrated with the best shape having a capacity of approximately 8000 p.c.u./h with both signal and offside priority control. It is interesting to note that the two worst shapes of widening from the point of view of capacity are parallel sided widening, which is usual with signal control, and a square layout which is usual with roundabout control.

The effect of variations in the internal layout is illustrated by figure 23.2, where it can be seen that, for the shape of widening that gave the highest capacity in figure 23.1, priority and signal control gave similar maximum capacities. This occurred with an asymmetrically tapered flare widening with a flare of 1 in 4, providing much more width on entry than on exit.

It was concluded that vehicles which start from a wide front on entry, can merge through a narrow exit without difficulty or congestion if the rate at which their passage is narrowed downstream from the stop line is no greater than the rate of widening up to that line. With the three best shapes it can be seen from figure 23.1 that there is very little difference between signal and priority control.

The track experiments also showed that the capacity of conventional roundabouts, which had a fairly large island and a relatively narrow circulating road, could be improved by reducing the diameter of the central island to about one-third of the diameter of the circle inscribed within the outer kerbline of the roundabout. If the roundabout island diameter were further reduced, and if there was no positive means of deflecting vehicles to the nearside, vehicles tended to become congested around the small roundabout. Causing drivers to fully observe the offside priority rule however, by deflecting vehicles to the nearside on entry to the roundabout, removed this congestion and allowed the central roundabout to be reduced to practically nothing.

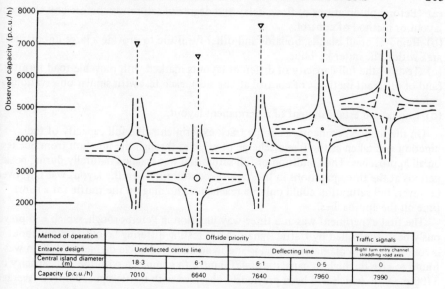

Method of operation	Offside priority				Traffic signals
Entrance design	Undeflected centre line		Deflecting line		Right turn entry channel straddling road axes
Central island diameter (m)	18·3	6·1	6·1	0·5	0
Capacity (p.c.u./h)	7010	6640	7640	7960	7990

Figure 23.2 Effect of variation of internal layout and method of operating in the best outline shape (based on ref. 1)

This means that the separation of conflicting traffic movements achieved by the normal roundabout island can be performed more efficiently by deflecting vehicles to the nearside on entry to the intersection.

These track experiments indicated that substantial improvements in junction capacity could be obtained by the use of modified layouts.

As a result of these experiments it was found that the highest capacities obtained in intersections of different sizes, regardless of shape, internal layout and method of control, could be related approximately to the dimensions by the following simple formula

$$q = k(\Sigma w + a^{1/2})$$

where q is the capacity in p.c.u./h,

k is an efficiency coefficient, which was found to be about 100 p.c.u./h per m for the highest capacities,

Σw is the sum of the basic road widths in metres used by traffic in both directions to and from the intersection,

a is the area of widening in square metres, that is the area within the intersection outline including islands, if any, lying outside the area of the basic crossroads.

At most practical intersections on the public highway the value of k can be expected to be less than 100.

Pedestrian effects, safety requirements and the reaction of drivers to new traffic situations make it necessary however to correlate test track experience with performance on the public highway. Three important junctions of three, four and five roads respectively were chosen and the following procedure was adopted.

(a) 'Before' observations of flow, delay and degree of saturation; before any change of layout or method of control.

(b) Removal of all islands, bollards and other furniture to provide a large unobstructed area within the outer kerbline.

(c) Tests of the full capacity of different layouts marked with movable road furniture (and of different methods of control at one site); each test with similar observations as before.

(d) Choice and establishment of a permanent layout.

On the road experiments, as in the track experiments, the full capacity of the junction was taken as the free discharge of vehicles through the junction from queues on all approaches. The condition of full saturation was reached naturally during peak periods at the three junctions in the 'before' condition. When the layout was improved however, full saturation could only be reached by holding up the traffic for a short time on the approaches.

The first experiment was at a three-way junction in Peterborough, which had previously been controlled by traffic signals. After testing differing layouts a roundabout layout with a 3 m diameter central island in an inscribed circle diameter of 29 m was found to give the highest capacity of 4700 p.c.u./h compared with the full capacity of 3700 p.c.u./h, which had been previously observed with traffic signal control. Average delay under the modified form of control was also found to be reduced to less than half at all levels of flow. Pedestrian delays were also found to be less with the new layout, although some initial difficulty was experienced before pedestrians became used to the change in the form of control.

The second public road experiment was at a five-way roundabout in Cardiff where an egg-shaped island, about 30 m across, was finally replaced by a small island of 5 m diameter within an inscribed circle of 46 m diameter. The capacity was found to be increased in this instance by nearly 25 per cent. A difficulty at this intersection was the number and type of the conflicting movements, which appeared to restrict the benefit that could be obtained from the extra area.

A third experiment was carried out at a four-way roundabout in Hillingdon, 20 km west of London, and on a main radial road. A wide variety of layouts were tested at the site including traffic signal control and roundabouts with different size central islands and alternative entrance layouts. The arrangement that gave the greatest increase in capacity, some 35 per cent, was a roundabout with a 15 m diameter central island with the outer kerbs having a diameter of 46 m. It was considered by the Transport and Road Research Laboratory that the relatively large central island for optimum capacity was caused by incorrect entrance deflection and wide approach roads.

Principles of design of small island roundabout intersections

As a result of the test track experiments and experience on public highways, the Department of the Environment issued recommendations for the design of small island roundabouts[2] and these were consolidated in recommendations for the general design roundabouts in Department of the Environment Technical Memorandum H2/75. [3]

The principles of the design of small island roundabouts are outlined in this memorandum. It is important, especially on faster roads, to reduce vehicle speeds to below 50 km/h. This is achieved by the deflection of vehicle paths so that straight through vehicle paths do not occur, by using a suitable size and position for the central

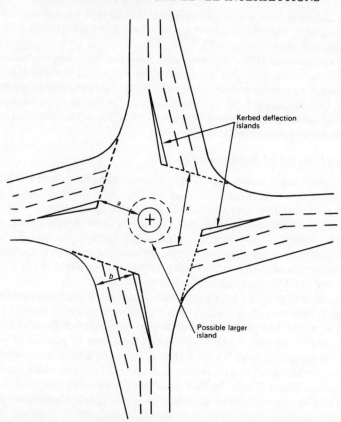

Figure 23.3 A four-way junction controlled by a small diameter roundabout with asymmetric widening and full deflection (based on ref. 2)

land, by the use of a stagger between entrance and exit, by the provision of angled deflection islands and carriageway markings in the entrances or by an adjustment of the alignment of exit and entrance so that they are not parallel.

The central islands are smaller than in conventional roundabouts but they must still be of sufficient size to deflect vehicle paths. An approximate rule is that the diameter of the central island should be one-third of the inscribed circle diameter or diameter of the circle which may be inscribed within the outer carriageway kerbline. It is also desirable to limit the width of the circulatory carriageway to 15 m. It may also be possible to employ smaller islands but speeds must be kept low by other means of deflection.

Entries should be flared so that with a single or 2 lane approach, 3 and 4 lanes are provided at the 'give way' line. The width of the traffic lanes should be between 2.5 and 3.5 m and the taper of the flare approximately 1 in 3. The 'give way' line should be sited at a distance from the central island approximately equal to the width of the preceding entry at its 'give way' line (dimensions a and b in figure 23.3). The size of the junction should also be adequate to provide a stopping distance from the 'give way' line to the point of conflict with a vehicle from the left of at least 25 m (dimension x in figure 23.3). The exits from a roundabout should also be tapered at about half the entr taper. A typical small island roundabout layout where speed reduction is achieved by deflection islands is shown in figure 23.3.

Limited experience on public highways with these junctions suggests that small island roundabouts are comparable from a safety aspect to conventional roundabouts especially when they replace priority or signal control. However, when they replace conventional roundabouts in built up areas an increase in accident rate has been noted. It is considered that this increase may be due to lack of adequate deflection resulting in excessive speed within the circulating area.

Principles of design of mini roundabouts

Where central islands are smaller than 4 m in diameter the roundabout is referred to as mini roundabout. The islands are not kerbed but flush with the road surface or slightly raised above the road. They may or may not have flared approaches.

Mini roundabouts are only recommended for use in Great Britain at existing junction in urban areas where the road layout and surrounding features prevent vehicle speeds exceeding 60 km/h on the intersection approaches and the junction is too constricted for large vehicles to manoeuvre without over-running the central island of a small roundabout and extra land acquisition is difficult.

Mini roundabouts normally have a maximum diameter of inscribed circle of approx mately 22 m; if greater space is available it is usually appropriate to provide a small roundabout with a kerbed central island of at least 4 m diameter. Because this type of roundabout is only provided at existing junctions it may not be possible to provide the same degree of deflection as with a small roundabout. The central island should be less than 4 m and greater than 1 m in diameter in the form of a dome with a maximum height not greater than 125 mm. So that the island can be easily perceived by approach drivers it should be painted with reflectorised or thermoplastic white paint. Approache should be marked into as many lanes as is possible, should normally be not less than 2.5 m at the 'give way' line but may be tapered to a minimum of 2 m on the approach

The capacity of both small and mini roundabouts is given by the formula

$$Q_p = k \left(\Sigma w + a^{1/2} \right) \tag{23.1}$$

where Q_p is the practical capacity in vehicles per hour; Σw is the sum of the basic roa widths (not half width) on all approaches (m); a is the area added to the junction by th flared approaches (m^2); k is a factor depending upon the type of roundabout, ranging from 70 for a 3 arm small roundabout to 40 for a 5 arm mini roundabout. This formu is applicable to a heavy vehicle content less than 15 per cent. For a heavy vehicle cont of 15 to 20 and 20 to 25 per cent the calculated value should be reduced by 4 and 8 per cent respectively. For design purposes 85 per cent of the calculated value of capac should be used.

References

1. F. C. Blackmore. Capacity of single level intersections. Transport and Road Research Laboratory Report L.R. 356 (1970)
2. Department of the Environment. Technical Memorandum H7/71. Junction Design
3. Department of the Environment. Technical Memorandum H2/75. Roundabout Design

Problems

Are the following statements correct or incorrect?

1. For a given shape of intersection and form of control, greater capacity can be obtained by increasing the plan area of the junction.

2. At a straight crossroad intersection under traffic signal control the most efficient manner in which the capacity can be increased is by parallel sided widening of the approaches.

3. At a roundabout intersection the least efficient road layout is formed by a square layout with the approach highways aligned along the diagonals.

4. The increased traffic capacity associated with mini-roundabouts is partially caused by the deflection to the nearside of entering vehicles.

5. It is possible to express the maximum capacity of an intersection under both signal and priority control by a single formula with a differing constant term.

6. A disadvantage of the mini-roundabout when compared with a conventional roundabout is the ease with which the mini-roundabout is blocked by abnormally large vehicles.

Solutions

1. This statement is correct. The capacity of any intersection under either signal or priority control may be increased by enlarging the area available for vehicle movement. This is illustrated in figure 23.1, where a straight crossroad intersection is widened from 31·8 m² to 1560 m². The increase in capacity for the same geometric shape when under signal control is from approximately 3700 p.c.u./h to approximately 6000 p.c.u./h.

2. This statement is incorrect. Reference to figure 23.1 shows that the parallel sided widening of a traffic signal-controlled intersection (maximum observed capacity approximately 6000 p.c.u./h) is far less efficient than a geometric design with tapered approaches, the most efficient of which has a maximum observed capacity of 8000 p.c.u./h.

3. This statement is correct. The use of the conventional square sided roundabout is much less efficient than other geometric shapes. This point is illustrated in figure 23.1 where the best form of geometric layout gives an increase in maximum capacity of approximately 14 per cent over the conventional roundabout shape.

4. This statement is correct. The deflection of vehicles entering the intersection to the nearside serves the same purpose as the larger central islands. Without the deflection, and with small central islands, vehicles tend to become congested in the centre of the intersection.

5. This statement is correct. The work on the test track of the Transport and Road Research Laboratory showed that maximum capacity of all geometric shapes of intersections under both priority and signal control was proportional to the sum of the basic road widths plus the square root of any widening area lying outside the basic intersection road outline.

6. This statement is incorrect. An advantage of the mini-roundabout is that the greater area of carriageway allows more space for abnormally large vehicles to negotiate the intersection.

24

Grade-separated junctions

Intersections between highways are a potential source of danger and congestion unless they are designed to have a similar capacity to the approach highways and allow smooth uninterrupted flow for heavy traffic movements.

In some circumstances, such as in urban areas, the consequences of not having smooth uninterrupted flow may not be serious enough to warrant the constructional costs of providing grade-separated intersections. The cost of vehicle delay at the single-level intersection is usually the major criterion that is then used to justify the provision of grade separation.

On high-speed routes however the dangers of an interruption in the flow of vehicles are so great that grade-separated junctions are justified even at low traffic volumes.

It has been stated[1] that grade separation is usually warranted on the following grounds:

(a) Sufficient traffic capacity cannot be obtained by the use of a single-level scheme;
(b) The construction of a grade-separated scheme is justified by the saving in delay and accidents compared with a single-level intersection;
(c) A grade-separated scheme is more economical than a single-level scheme because of the topography of the site or because of existing development;
(d) Because of the need to maintain similar junction types on a highway or maintain high speeds and uninterrupted flow on a major route such as a motorway.

The form of the grade separation will vary according to the importance of the differing traffic movements and designs will usually be justified by the criteria already stated. While a considerable number of design variants are obviously possible, a few of the common standard forms of grade-separated intersections will now be described.

Three-way junctions

At a three-way junction the usual layout adopted is the trumpet, illustrated in figure 24.1. It allows a full range of turning movements but suffers from the disadvantage that there is a speed limitation on the minor road right-turning flow due to radius size. Where topographical conditions make it necessary the junction can be designed to the opposite hand, as shown in figure 24.2, but this layout has the disadvantage that vehicles leaving the major road have to turn through a small radius.

219

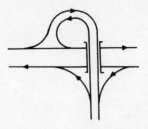

Figure 24.1 The usual form of trumpet intersection

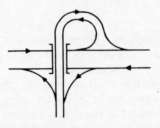

Figure 24.2 A form of trumpet intersection to avoid existing development

Where a limited range of turning movements is required then one of the junctions of the form shown in figure 24.3 may be used. The layout shown in (c) may be used in a situation in which the two carriageways are a considerable distance apart and where a right turn exit on the right-hand lane would not be inappropriate.

It may sometimes in the future become necessary to convert a three-way junction to a four-way junction, and the design should allow for this future conversion. A suitable form of junction is the partial bridged rotary shown in figure 24.4. Alternatively the trumpet intersection shown in figure 24.1 may be converted to a cloverleaf intersection

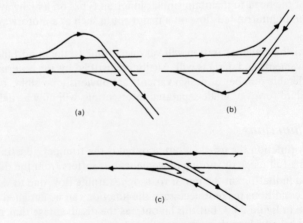

Figure 24.3 Alternative forms of three-way junctions

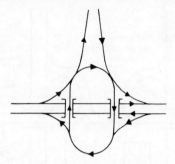

Figure 24.4 A partial bridged rotary intersection

Four-way junctions

Frequently four-way junctions are formed by the intersection of a major and a minor road. In such instances it is often possible to allow traffic conflicts to take place on the minor road. The simplest type of intersection where this occurs is the diamond, shown in figure 24.5. This is a suitable form of layout in which there are relatively few turning movements from the major road onto the minor road since the capacity of the two exit slip roads from the major road is limited by the capacity of the priority intersection with the minor road. Care needs to be taken with this and other designs by the use of channelisation, so that wrong-way movements on the slip roads are prevented.

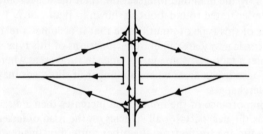

Figure 24.5 A diamond intersection

In some circumstances site conditions may prevent the construction of slip roads in all of the four quadrants as required by the diamond. A solution in this case would be the construction of a half cloverleaf as shown in figure 24.6(a). Where site conditions make it necessary or where right-turning movements from the major road are heavy then the opposite hand arrangement as shown in figure 24.6(b) may be used, with the disadvantage however of a more sudden speed reduction for traffic leaving the major road. As the traffic importance of the minor road and the magnitude of the turning movements increase additional slip roads may be inserted into the partial cloverleaf design until the junction layout approaches the full cloverleaf design.

Where major traffic routes intersect, it is no longer possible for the traffic conflicts to be resolved on the road of lesser traffic importance and the simpler junction types previously described are then likely to be unsatisfactory, both from the point of view of capacity and accident potential. The design of intersections of this type is always

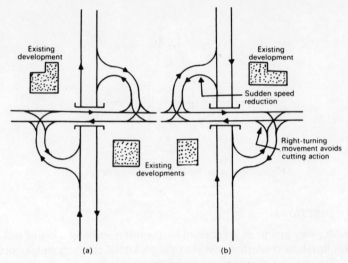

Figure 24.6 Partial cloverleaf designs

carried out after consideration of the individual directional traffic flows, having regard
for the topographical features of the area. Nevertheless it is possible to discuss the
general forms of these intersections and the traffic flows that are likely to warrant their
construction. When turning movements from the route of greater traffic importance
cannot be handled by the diamond intersection, then the usual solution in Great Britain
has been the grade-separated roundabout as shown in figure 24.7.

The advantages of this type of junction are that it occupies a relatively small area of
land and has less carriageway area than other junctions of this type. It also allows easy
U-turns to be made, a factor of some importance in rural areas where intersections may
be widely spaced or in urban areas where a number of slip roads may join the major
route between interchanges.

As the traffic importance of the minor road increases then a disadvantage of this
form of junction is the necessity for all vehicles on the road of lesser traffic importance
to weave with the turning traffic from the other route. For this reason the capacity of
the weaving sections limits the capacity of the intersection as a whole.

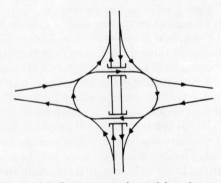

Figure 24.7 Grade-separated roundabout intersection

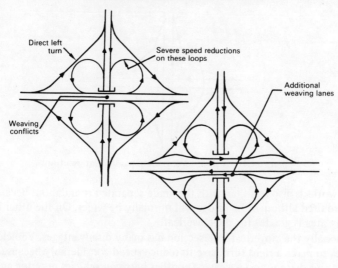

Figure 24.8 Cloverleaf intersections

An alternative form of intersection which, while popular in the United States, has found limited application in Great Britain is the cloverleaf, shown in figure 24.8. With this form the straight ahead traffic on both routes is unimpeded and in addition left-turning movements may be made directly from one route to the other. The cost of

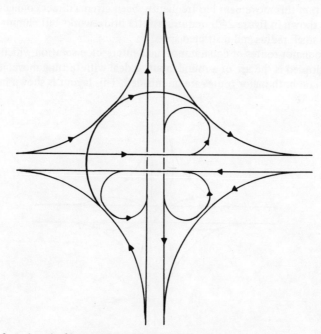

Figure 24.9 A cloverleaf intersection with a direct link for a heavy right-turning movement

Figure 24.10 A three-level grade-separated intersection

structural works is also less than with the grade separated roundabout because only one bridge is required although the bridge will normally be wider. On the other hand the carriageway area is greater for the cloverleaf.

Operationally the cloverleaf intersection has many disadvantages. Vehicles leaving both routes to make a right turn have to reduce speed considerably because of radius restrictions and there is also a weaving conflict between vehicles entering and leaving each route, which peak at the same time. This conflict can however be reduced by the provision of a separate weaving lane on either side of each carriageway but this will increase carriageway and bridge costs. There are also four traffic connections onto each route and U-turns may present some difficulty to drivers unfamiliar with the junction. Sign posting has also been stated to present some difficulties.

Where heavy right-turning movements in one direction are anticipated at a cloverleaf type intersection this movement has frequently been given a direct connection by a single link as shown in figure 24.9, requiring extra bridgeworks but eliminating one loop with its small radius and restricted speed.

Where two major routes of equal importance intersect, a solution which has been frequently adopted is the use of a roundabout to deal with turning movements while straight flows on both major routes are unimpeded. This layout is shown in figure 24.1(

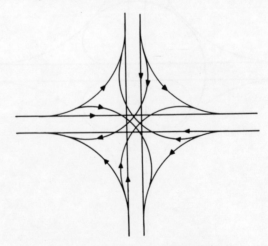

Figure 24.11 Direct connection between two primary distributors or motorways

Department of the Environment Technical Memorandum H6/75 relating to the design of Rural Motorway to Motorway Interchanges states that this type of non-free-flow junction may be appropriate when the major traffic movements are straight ahead, those turning are relatively low in volume and not predominantly right turners or heavies. When there are four dual three-lane approaches, however, the turning flows should be carefully examined to ensure that in the design year hesitation and yielding but not queuing take place on the roundabout approaches. If there are heavy single-turning movements then an independent direct link or a left-hand filter lane may be used with this type of junction to prevent queueing.

An alternative form of intersection when turning flows are high and well balanced, especially if the flow contains upwards of 20 per cent of heavy vehicles, is the multi-level free-flow layout shown in figure 24.11.

Reference

1. Layout of roads in rural areas. Ministry of Transport. H.M.S.O. (1968)

Problems

From the following junction types (a)–(i) select a layout that is likely to be satisfactory for the site and traffic conditions described in 1–6.

(a) A grade-separated rotary intersection with all the minor road flow passing around the rotary section as shown in figure 24.12(a).

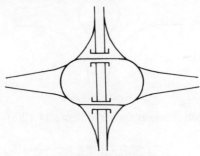

Figure 24.12(a)

(b) A grade-separated rotary intersection with only turning movements passing around the rotary section as shown in figure 24.12(b).

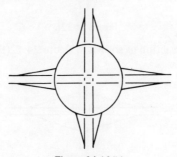

Figure 24.12(b)

(c) A single-level four-way signal-controlled intersection as shown in figure 24.12(c).

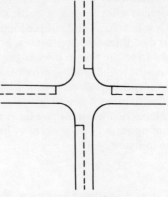

Figure 24.12(c)

(d) A single-level four-way roundabout intersection as shown in figure 24.12(d).

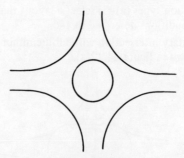

Figure 24.12(d)

(e) A semi-cloverleaf intersection as shown in figure 24.12(e).

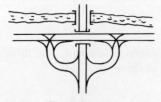

Figure 24.12(e)

(f) A diamond-type intersection as shown in figure 24.12(f).

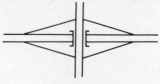

Figure 24.12(f)

(g) A trumpet intersection as shown in figure 24.12(g).

Figure 24.12(g)

(h) A semi-grade-separated roundabout as shown in figure 24.12(h).

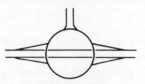

Figure 24.12(h)

(i) A grade-separated intersection with direct connections for all movements as shown in figure 24.12(i).

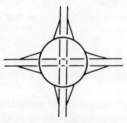

Figure 24.12(i)

1. Two major rural highways intersect at a right angled T-junction with heavy right turning movements from both highways. The traffic flows and accident potential are so great that grade separation is desirable.

2. In an urban area a heavily trafficked principal traffic route crosses a less heavily trafficked highway and there are a considerable number of turning movements between the highways.

3. Two motorways intersect in a rural area and it is desired to allow for a full range of turning movements between the two routes.

4. An all-purpose highway is crossed by a motorway but it is not anticipated that there will be many turning movements between the two highways. There are no topographical limitations on the design of the junction.

5. The junction is as described in (4) but a river runs parallel with the all-purpose highway preventing any entry or exit highways on one side of the all-purpose highway.

6. A right angled T-junction is to be constructed between a motorway and an all-purpose highway and it is anticipated that in the future the all-purpose highway will be extended so that a four-way connection will be required.

Solutions

The junction layouts that would be chosen for the given situation are:

1. At this intersection grade separation is desirable and as future extension of the three-way layout is not considered, then the trumpet intersection (g) would appear suitable.

2. As this intersection is heavily trafficked and there is a considerable number of turning movements then grade separation is desirable. It is an urban situation where lane is likely to be limited and so a grade-separated roundabout (a) with all the traffic from the less heavily trafficked route passing around the roundabout would be used.

3. At the intersection of two motorways where provision for all turning movements has to be made then it is usual in Great Britain to use a grade-separated rotary intersection (b) where a relatively compact layout is required or an intersection with direct connections (i) where more space is available or where speed reductions for the turning movements are undesirable.

4. Where an all-purpose highway crosses a motorway and there are few turning movements and no topographical limitations then a diamond intersection (f) is usual.

5. Where a diamond intersection cannot be provided because of physical obstructions then a semi-cloverleaf intersection (e) allows all the slip roads to be provided on one side of the minor road.

6. At a T-junction between a motorway and an all-purpose highway where it is required in the future to convert the T-junction into a four-way junction then a semi-grade separated roundabout (h) may be preferred. This layout allows the easy extension of the minor road into a four-way junction.

25

The environmental effects of highway traffic noise

Noise; measurement of sound levels

In our industrial society the number of sources of sound are steadily increasing and when these sounds become unwanted they may be classed as noises. Sound is propagated as a pressure wave and so an obvious measure of sound levels is the pressure fluctuation imposed above the ambient pressure.

If the graph of pressure against time for a single frequency is examined it is found to have a maximum amplitude (P_m) and in sound pressure measurements it is the root-mean-square pressure that is recorded. Some sounds are a combination of many frequencies while others are composed of a continuous distribution of frequencies. When this occurs the root-mean-square pressure values of all the individual frequencies are added together.

Using pressure units to describe sound levels requires a considerable range of numbers. It is frequently stated that the quietest sound that most people can hear has a sound pressure level of approximately 20 micropascals (μN m^{-2}) while at 100 m away from a Saturn rocket on take-off the sound pressure level is approximately 200 kPa. Rather than use a measurement system with this range, the ratio of a sound pressure to a reference pressure is used so that the sound pressure level is given in decibels by the ratio

$$20 \log_{10} \frac{\text{pressure measured}}{\text{reference pressure}} \text{ decibel (dB)}$$

The reference pressure is taken as 20 μPa.

Some idea of the range of sound pressure levels measured in decibels can be obtained from the values given in table 25.1.

When sound pressure levels are measured adjacent to a highway, a meter measuring in dB might indicate the same value when a fast moving motor cycle with a high pitched or high-frequency engine note passes and when a slow moving goods vehicle passes with a lower frequency note. The reason why the high pitched note is usually found more annoying than the lower one from the goods vehicle is that the human ear is more sensitive to sounds with higher frequencies than it is to sounds with lower frequencies.

229

TABLE 25.1 Some typical sound pressure levels expressed in dB

Sound	Approximate sound pressure level (dB)
pneumatic drill	120
busy street	90
normal conversation	60
quiet office	50
library	40
quiet conversation	30
quiet church	20

If a sound level scale is going to be useful for measuring annoyance to human beings it must take this effect into account. Such a scale is measured in dB(A), the sound level measurements then being obtained by an instrument that weights the differing frequenc components according to the curve given in figure 25.1. This results in those frequencie

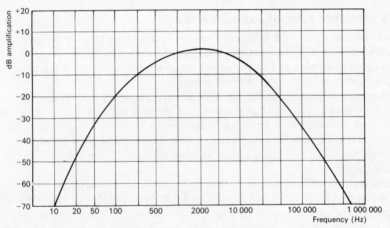

Figure 25.1 The 'A' weighting curve for sound-level meters

that are relatively high or low receiving less weighting than those in the range 1 to 4 kH: With a sound level meter reading in dB(A) it would therefore be found that the higher frequency note of the fast motor cycle would give a higher reading in dB(A) than the goods vehicle, although both produce the same sound pressure level measured in dB.

Sometimes it is necessary to know not only the sound level in dB or dB(A) but also the contribution that differing frequencies make to the overall sound. To obtain this information the sound is analysed by an instrument that passes it through a system of filters and allows the relative proportions to be determined.

Road traffic noise differs from most other sources of noise in that the level of noise varies both considerably and rapidly. If the variations of sound pressure level with time are recorded then a record of the type shown in figure 25.2 is obtained. At low sound-pressure levels the noise emitted from vehicles does not cause a great deal of annoyance but at higher levels the annoyance is considerable. For this reason many measures of noise nuisance specify a sound pressure level that is exceeded for 10, 20, 30 per cent etc. of the time.

When considering a scale that can be used to express a level of noise that should not be exceeded, it should be noted that this scale should be capable of expressing the relative effect on people of the noise being measured. Scholes and Sargent[1] have stated that the unit selected should meet at least the following requirements. Firstly the unit should correlate reasonably well with the criterion of dissatisfaction chosen so that noise levels measured using the unit will represent subjective reactions. Secondly a reasonably accurate set of design rules should be available, covering the estimation of noise exposure from traffic data and the estimation of the performance of noise control techniques, in terms of the chosen unit.

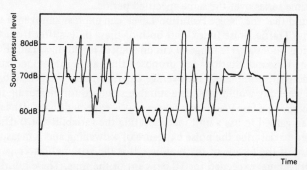

Figure 25.2 The variation of sound-pressure level with time for highway traffic

The London Noise Survey[2] measured noise levels at 540 sites and the subjective reactions of 1300 residents. During this survey particular values noted were the L_{10} level (the sound pressure level in dB(A) exceeded for 10 per cent of the time) and the L_{90} level (the sound pressure level in dB(A) exceeded for 90 per cent of the time). These levels represent the extremes of the range that were recorded for 80 per cent of the time. These values are referred to as the 'noise climate' and, together with the L_{10} value, are often quoted in the Wilson Committee Report[3].

Using data from interviews with 1200 residents at 14 sites in the London area where the roads were all straight, level and carrying free-flowing traffic, Langdon and Scholes[4] developed the 'Traffic Noise Index' (TNI). They found this index correlated well with dissatisfaction with noise conditions when the TNI was given by

$$4(L_{10} - L_{90}) + L_{90} - 30$$

A study carried out in Sweden[5] showed a good correlation between noise disturbance and three measures of noise, L_{10}, L_{50} and a noise exposure index based on the energy mean of the noise level, and given by the expression

$$L_{eq} = K \log \frac{1}{100} \Sigma 10^{L_i/K} f_i$$

where K is an empirically determined constant,
$\quad L_i$ is the median sound level for the 5 dBA interval i
$\quad f_i$ is the percentage time that a sound level is in the ith interval.

This unit has not however been found to correlate well with experience in the U.K. The difference in performance between the two countries is considered to be caused by the lower noise levels and the greater variability experienced in Swedish traffic conditions.

Another unit has been proposed[6] to cover a range of noise sources, whether highway traffic noise, aircraft noise or laboratory noise. It is referred to as the 'Noise Pollution Level' (LNP) and is given by the following expression

$$LNP = L_{eq} + 2 \cdot 56\sigma$$

where L_{eq} is the energy mean noise level of a specified period
σ is the standard deviation of the instantaneous sound level considered as a statistical time series over the same specified period.

It has been found that the 'Noise Pollution Level' can express annoyance with traffic noise as well as the 'Traffic Noise Index' but both of these units suffer from the fact that their prediction under a wide range of circumstances is still uncertain. For this reason the Building Research Station[1] has proposed that, as an interim measure until further investigations are completed, the average L_{10} taken over the period 6 a.m. to 12 midnight on a weekday would provide a suitable standard for measuring traffic noise nuisance in dwellings. In a few years time however it is expected that sufficient experience will have been gained to use a unit incorporating the variability of traffic noise.

This unit is used to describe the noise exposure of a dwelling and is measured at 1 m from the mid-point of the facade of the building. It is the arithmetic average of the hourly levels in dB(A) just exceeded for 10 per cent of the time. These hourly values are obtained from sampling. The duration of each sample should include the passage of at least 50 vehicles and preferably 100 vehicles.

A maximum acceptable level of noise nuisance

The setting of an acceptable or a maximum level of noise presents considerable difficulties in compromising between what is desirable and what is physically and economically possible. The Wilson Committee[3] made a number of recommendations in 1963 in terms of L_{10} levels averaged over a 24-hour period. The maximum L_{10} levels inside buildings recommended by this Committee are given in table 25.2.

TABLE 25.2   Wilson Committee recommendation for
maximum L_{10} levels indoors

Situation	Day	Night
country areas	40 dBA	30 dBA
suburban areas	45 dBA	35 dBA
busy urban areas	50 dBA	35 dBA

Scholes and Sargent[1] have suggested that as an interim standard a value of L_{10} (6 a.m.–midnight) of 70 dB(A) at residential facades should not be exceeded. The Noise Advisory Council has also recommended that as an act of conscious public policy existing residential development should in no circumstances be subjected to a noise level of more than 70 dB(A) on the L_{10} index. This view was also shared by the Urban Motorways Committee in their report 'New Roads in Towns'[8].

The prediction of noise levels

The major factors which influence the generation of road traffic noise are:

(a) the traffic flow;
(b) the traffic speed;
(c) the proportion of heavy vehicles;
(d) the gradient of the road;
(e) the nature of the road surface.

In addition the following factors influence the noise level at a reception point distant from the highway:

(f) attenuation of the sound waves due to distance between source and receiver and also due to ground absorption;
(g) obstruction of the sound waves by buildings or noise barriers;
(h) obstruction of the sound waves due to a restricted angle of view of the source line from the reception point;
(i) reflection effects.

When predicting traffic noise levels by the procedure given in the Department of the Environment Memorandum, Calculation of Road Traffic Noise the factors (a)–(e) are used to predict the basic noise level (in terms of the hourly L_{10} or the 18-hour L_{10}) and the factors (f)–(i) are used to modify the basic noise level to obtain the prevailing noise level at a reception point.

Initially the basic noise level in terms of the 18-hour L_{10} or hourly L_{10} noise level is determined by the 18-hour or hourly traffic flow for a normalised source to receiver distance of 10 m when the mean traffic stream speed is 75 km/h, there are no heavy vehicles in the flow and the roadway is level. The traffic flow (Q) is the two-way flow from 06.00 h to 24.00 h or the hourly flow (q) except when the two carriageways are separated by more than 5 m or where the heights of the outer edges of the two carriageways differ by more than 1 m. In these cases the noise level of each carriageway is evaluated separately.

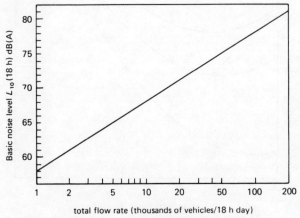

Figure 25.3 Basic noise level L_{10} (18 h)

The relationship between L_{10} (18 h) basic noise level and traffic volume is shown graphically in figure 25.3 and expressed mathematically as

$$L_{10} \text{ (18 h)} = 28 \cdot 1 + 10 \log Q \text{ dB(A)}$$

also $\qquad L_{10} \text{ (hourly)} = 41 \cdot 2 + 10 \log q \text{ dB(A)}.$

A correction has to be made for the mean traffic stream speed and the percentage of heavy vehicles where these differ from 75 km/h and zero per cent respectively. The correction is shown in figure 25.4. Mathematically this correction is given by:

$$\text{Correction} = 33 \log \left(V + 40 + \frac{500}{V} \right) + 10 \log \left(1 + \frac{5P}{V} \right) - 68 \cdot 8 \text{ dB(A)}$$

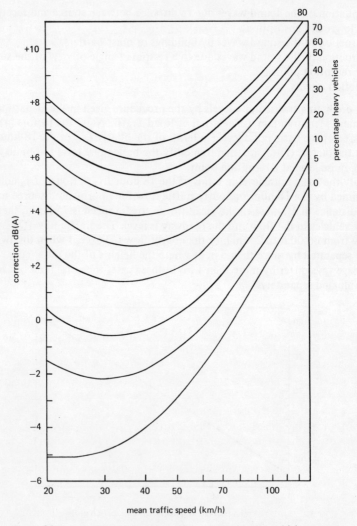

Figure 25.4 Correction for mean traffic speed and heavy vehicle content

where P is the percentage of heavy vehicles in the flow (a heavy vehicle is any vehicle, other than a motor car, the unladen weight of which exceeds 1525 kg).

The traffic speed (V) may be obtained in either of two ways. It may be the prescribed highest mean speed in any one year within a 15-year period as given in table 25.3, or where local conditions indicate a significantly different value from the prescribed mean speed, then the highway authority may estimate the highest mean speed in a 15-year period.

Alternatively when the prevailing noise level is being calculated then an estimate or actual measurement of existing speed is to be used.

An additional correction has to be made for the extra noise generated by traffic on a gradient. When the prevailing or existing noise is being calculated then the correction for gradient effect is given by a correction of 0·3 x the percentage gradient dB(A). When future traffic noise levels are being calculated and a prescribed speed is used then the correction is 0·2 x the percentage gradient dB(A). The lower value is used in this case because an increase in speed has been taken into account. Where carriageways are

TABLE 25.3   Prescribed highest mean speeds in any one year within 15 years

Highway type		Prescribed speed
Special roads (rural) excluding slip roads	(speed limit less than 60 mph)	108 km/h
Special roads (urban) excluding slip roads	"	97 km/h
All-purpose dual carriageways excluding slip roads	"	97 km/h
Single carriageways, more than 9 m wide	"	88 km/h
Single carriageways, 9 m wide or less	"	81 km/h
(Slip roads are to be estimated individually)	"	
Dual carriageways	(speed limit 50 mph)	80 km/h
Single carriageways	"	70 km/h
Dual carriageways	(speed limit < 50 mph > 30 mph)	60 km/h
Single carriageways	"	50 km/h
All carriageways	subject to a speed limit of 30 mph or less	50 km/h

separated by more than 5 m or where there is one-way traffic then the correction only applies for the upward flow. If there is a single direction down gradient the Memorandum recommends the use of sound level measurements.

A final adjustment to the basic noise level must then be made for the effect of road surface texture. If deep random grooving is present (grooves with a depth of 5 mm or more) then tyre/road interaction produces an increase in noise level given by $4 - 0·03p$ dB(A) where (p) is the percentage of heavy vehicles.

With the basic noise level determined it is necessary to make corrections for the factors which effect the propagation of sound between the source and the reception position.

First a correction is made for the distance between the source and reception position. The nature of the ground influences distance attenuation and the Memorandum divides the ground over which the sound is propagated into hard ground and grassland. Hard ground is defined as mainly level ground, the surface of which is predominantly (more than 50 per cent) non-absorbent, that is, paved, concrete, bituminous surfaces and water.

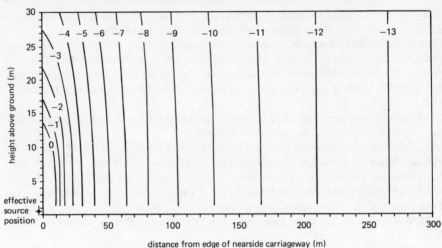

Figure 25.5 Correction for propagation over hard ground

The Memorandum gives the distance attenuation in these circumstances as:

$$\text{distance correction} = -10 \log (d'/13 \cdot 5) \text{ dB(A)}$$

where d' = the minimum slant distance between the effective source position and the reception point.

The effective source line is assumed to be $3 \cdot 5$ m in from the near kerb and at a height of $0 \cdot 5$ m above the road surface. Where the carriageways are considered separately then the source line for the far carriageway is taken as $3 \cdot 5$ m in from the far kerb and the distance from the kerb to be used in figures 25.5 and 25.6 is taken as 7 m in from the far kerb.

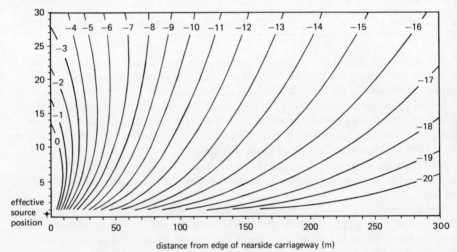

Figure 25.6 Correction for propagation over grassland

The distance correction is obtained graphically from figure 25.5 and in this chart the distance is measured from the edge of the nearside carriageway.

Where the surface between the source line of the noise and the reception point is predominantly of an absorbent nature, such as grass, cultivated or planted land, the effects of ground absorption must be considered in addition to distance attenuation. Mathematically the correction is given as:

$$\text{Correction} = -10 \log (d'/13 \cdot 5) + 5 \cdot 2 \log [3h/(d + 3 \cdot 5)] \text{ dB(A)}$$

$$1 \leqslant h \leqslant (d + 3 \cdot 5)/3$$

$$= -10 \log (d'/13 \cdot 5) \text{ dB(A)}$$

$$h > (d + 3 \cdot 5)/3$$

where d' is the slant distance between source line and reception point and d is the horizontal distance between the edge of the nearside carriageway and the reception point.

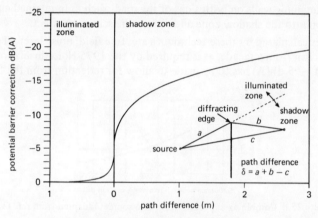

Figure 25.7 Potential barrier correction

This distance and ground absorption correction is obtained graphically from figure 25.6 and in this chart the distance is measured from the nearside kerb or as previously stated for roads where the two carriageways are being considered separately.

In many practical cases the screening effect of objects between the source line and receiver must be taken into account. Dealing initially with the attenuation due to either a long noise barrier or a continuous obstruction caused by site conditions. The basis of the correction is the path difference between the source line and the reception point as illustrated in figure 25.7.

This correction is applied to the basic noise level which has been corrected for distance using the hard ground correction. Ground absorption is ignored since the near ground rays are obstructed by the barrier.

Where only part of the road is shielded, for example by a short barrier, then a modified correction procedure has to be used which is conveniently illustrated by the following steps:

a) Let θ_H be the total angle of view of the unscreened section of the source line for which the ground between the road and the reception point is hard. Let θ_S be the similar angle of view for which the ground is soft (grassland). Let θ_B be the total angle of view obstructed by barriers. (Normally for straight roads $\theta_B + \theta_H + \theta_S = 180°$)

(b) Calculate the contribution to the noise at the reception point due to those lengths of line source covered by θ_H using the hard ground propagation correction. Correct this for the restricted angle of view by applying the correction $10\log(\theta_H/180)$ dB(A).

(c) Repeat using θ_S and the grassland propagation and absorption correction and restricted field of view correction to obtain this contribution at the reception point. Combine with the value calculated in (b) to obtain L_U.

(d) Calculate the contribution from the screened section of road by calculating the unobstructed noise level at the reception point for hard ground and apply the long barrier correction and the restricted angle of view correction to give L_B.

(e) Combine L_U and L_B to give L_{10}.

In many situations reflection of noise from noise barriers or substantial buildings beyond the traffic stream along the opposite side of the road increases the noise level by 1 dB(A). Retaining walls on both sides of the road, such as in retained cut, also cause reflection into the shadow zone and reduce the screening effect.

These values calculated by these techniques are 'free field' noise levels. To calculate the noise level 1 m from a facade, as is required by the 1975 Noise Insulation Regulation a correction of +2·5 dB(A) has to be made to allow for reflection from the facade.

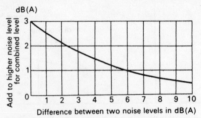

Figure 25.8 Combining exposures from two sources (adapted from ref. 1)

Controlling traffic noise by means of screens

When noise exposures from highway traffic are predicted it is frequently found that the predicted exposures are greater than the recommended maximum values. A method of reducing the noise exposure is by means of screens, making possible a reduction in nois exposure averaging 10 dB(A).

A disadvantage of screening is the size of the barriers required because if noise attenuation is to be obtained the facade being protected must be well within the sound shadow formed by the screen. It should not be possible to see over the top or around the ends of the screen if effective insulation is to be achieved. Narrow belts of trees and shrubs are considered relatively unsatisfactory by the Building Research Station for noise attenuation purposes. To be effective the belt should be about 50 m wide, dense and extend to ground level. The foliage should also be evergreen for all-year screening.

To produce an effective sound shadow a noise barrier should either be close to the highway or the building facade being protected. It should be dense enough to create an effective shadow—a recommended value is at least 10 kg/m²—and there should not be any sound paths either through or under the barrier. Structural and aesthetic considerations are also very important and these latter factors will obviously influence the design

Insulation of dwellings does however offer greater possibilities for the reduction of noise nuisance than can usually be obtained in existing situations by distance or noise barriers. It does not of course reduce noise nuisance in gardens and adjacent areas.

The use of thicker glass has only a marginal insulating effect (approximately 1-3 dB(A) improvement) but double windows with a space of at least 150 mm between the panes will give a sound insulation of between 20 and 30 dB(A) when one leaf is sealed and the other is well fitting. Unfortunately the insulation value is reduced when a powered fan is used to provide ventilation.

Controlling vehicle noise

The obvious way in which traffic noise may be reduced is by a reduction in the noise emitted by vehicles, but this solution presents many difficulties.

At the present time maximum permitted sound levels with which motor manufacturers in the United Kingdom must comply are fixed by Regulation 23 of the Motor Vehicles (Construction and Use) Regulations. Some typical limits are given in table 25.4. For vehicles first used on or after 1st November 1970 roadside limits are 3 dB(A) above those given in table 25.4.

TABLE 25.4   Maximum permitted sound levels in the United Kingdom

Description of vehicle	Maximum sound level dB(A)
motor cycle with a cylinder capacity exceeding 125 cc	86
goods vehicle with a maximum gross weight exceeding $3\frac{1}{2}$ tons	89
passenger vehicles constructed to carry not more than 12 passengers, exclusive of the driver	84

Measurement of individual vehicle noise at the roadside is not easy because in the United Kingdom the regulations require the measurement to be made at an open site to prevent reflections affecting the measurement and a noise background 10 dB below the vehicle peak. It is often difficult to find an open site where the traffic flow is low enough to meet the background noise level regulations and at the same time prevent the sound level equipment from being so obvious that offending drivers remove their foot from the accelerator.

Factors that affect the noise emitted by vehicles are speed, load, engine size, engine design and degree of enclosure, power unit inlet and exhaust noises, fan noise, diesel engine scavenging-blower noise, transmission noise, aerodynamic noise, road surface noise, tyre noise and brake noise.

While a detailed discussion of these noise effects, some of which are not well understood, is beyond the scope of this work, it is of interest to note that Burt[7] considers there should not be any serious technical or economic difficulty in reducing the levels for most cars from 84 dB(A) to 80 dB(A) by 1975. For the current type of commercial vehicle however the lowest predictable target over the next five years is probably a reduction from the present 89 dB(A) to 86 dB(A), since many commercial vehicles are now only marginally meeting the 89 dB(A) standard. For the heavier and more powerful commercial vehicle now being developed even the maintenance of 89 dB(A) is considered to be a significant development task, since current trends dictated by such factors as low torque, high-speed engines and the reduction of atmospheric pollution are not necessarily consistent with noise reduction. Some measure of the problem with commercial vehicles can be gained from the fact that at speeds in the region of 80 km/h coasting noise can reach 85 dB(A) and a reduction in noise will thus require attention to both vehicle design and the interaction between the vehicle and the road surface.

References

1. W. E. Scholes and J. W. Sargent. Designing against noise from road traffic. *Bldg Res. Stn Current Paper* CP.20/71
2. A. G. McKennel and E. A. Hunt. Noise annoyance in Central London. The Government Social Survey, *C.O.I. Report* SS.332 (1962)
3. Sir Alan Wilson, Chairman. Noise—Final report of the committee on the problem of noise. Cmnd 2056, London (1963)
4. F. J. Langdon and W. E. Scholes. The traffic noise index: a method of controlling noise nuisance. *Bldg Res. Stn Current Paper* CP.38/68
5. Statens Institut For Byggnasforskning. Trafikbuller i Bosladsomraden, *Statens Instit. For Byggnasforskning, Rapport* 36/68, Stockholm (1968)
6. D. W. Robinson. An outline guide to criteria for the limitation of urban noise. *Ministry of Technology, NPL Aero Report Ac. 39* (1969)
7. M. E. Burt. Roads and the environment. *Transp. Rd Res. Laboratory Report* 441 (1972)
8. Urban Motorways Committee. New roads in towns. Department of the Environment H.M.S.O. (1972)

Problems

1. The sound pressure in the driving cab of a heavy goods vehicle is 2 Pa. Is the sound pressure level measured in dB, with reference to 20 μPa: 80 dB, 100 dB or 150 dB?

2. A heavy goods vehicle travelling at 60 km/h is found to produce a noise with a very low frequency in the range 100–500 Hz while a high-performance car travelling at 120 km/h is found to produce a noise with a frequency in the range 1000–2000 Hz. If both vehicles produce a noise with the same pressure level, which noise source will register the higher noise pressure level when measured in dB(A)?

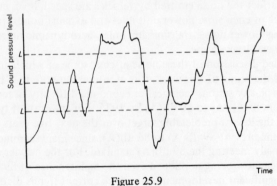

Figure 25.9

3. Variations of the sound pressure level with time at a site adjacent to the highway are shown in figure 25.9.

Indicate on the diagram which of the broken horizontal lines is likely to represent the L_{10}, L_{50} and L_{90} sound pressure levels.

4. Many indices of traffic noise have been developed. From the given list of indices select the appropriate index to meet the requirements listed.

Indices

(a) The traffic noise index;
(b) The sound pressure level in dB(A) exceeded for ten per cent of the time;
(c) The noise pollution level;
(d) The sound pressure level in dB(A) exceeded for ninety per cent of the time.

Requirements

(a) An index that has been extensively correlated with the dissatisfaction of human beings with traffic noise;
(b) An index that can be used to express the noise nuisance of both aircraft and highway traffic;
(c) An index that has limited correlation with dissatisfaction with highway traffic noise but which expresses the variability of sound pressure levels;
(d) An index that expresses the background noise level.

5. It has been suggested that the L_{10} noise index should not exceed a given value at building facades. Is this value in the region of 30 dBA, 70 dBA or 100 dBA?

6. Calculate the noise exposure level at 1 m from the facade of a building 45 m from the nearest edge of a free-flowing traffic lane. The average commercial vehicle content of the traffic stream is 40 per cent; the mean stream speed is 60 km/h; the traffic volume is 15 000 vehicles/day. A second free-flowing traffic stream is 50 m distant from the building facade; there are no commercial vehicles in this stream; the mean stream speed is 80 km/h; the traffic volume is 30 000 vehicles/day. Ignore the effect of ground attenuation and gradient.

7. Calculate the improvement in noise exposure that will be obtained by the use of a very long barrier adjacent to a motorway where the geometrical layout is as shown in figure 25.10.

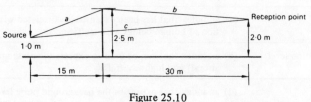

Figure 25.10

8. Calculate the improvement in noise exposure that will be obtained in the previous example if the barrier has a length of 200 m downstream and 50 m upstream of the reception point at which the noise exposure is to be calculated (see figure 25.11).

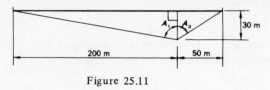

Figure 25.11

Solutions

1. Sound pressure level $= 20 \log_{10} \dfrac{\text{pressure measured}}{\text{reference pressure}}$ dB

$$= 20 \log_{10} \frac{2}{2 \times 10^{-5}} \text{ dB}$$

$$= 20 \times 5 \text{ dB}$$

$$= 100 \text{ dB}$$

2. Reference to figure 25.1 shows that sounds with frequencies in the range 100–500 Hz produce less effect on the human ear than sounds in the frequency range 1000–2000 Hz. If both sounds have the same sound pressure level when measured in dB then the high-performance car with a noise in the frequency range 1000–2000 Hz will register a higher sound pressure level in dB(A).

3. The correct placings of the L_{10}, L_{50} and L_{90} sound pressure levels are marked in figure 25.12.

4. The correct combination of indices and requirements is shown in table 25.5.

TABLE 25.5

Index	Requirement
(a) traffic noise index	(c) an index that has limited correlation with dissatisfaction with highway traffic noise but expresses the variability of sound pressure levels
(b) L_{10}	(a) an index that has been extensively correlated with the dissatisfaction of human beings with traffic noise
(c) noise pollution level	(b) an index that can be used to express the noise nuisance of both aircraft and highway traffic
(d) L_{90}	(d) an index that expresses the background noise level

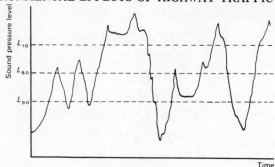

Figure 25.12

5. The Noise Advisory Council has recently recommended that the L_{10} index should be used for measuring traffic noise disturbance. It also recommended that as a conscious act of public policy existing residential development should not be subjected to a noise exposure level greater than 70 dB(A) on the L_{10} index unless remedial or compensatory action was taken by the responsible authority (Reported in Hansard, 24th June 1971).

6. Calculation of noise exposure level from stream 1.

From figure 25.3   For a flow rate of 15 000 vehicles/18 h day L_{10} (18 h) = 69·9 dB(A).
From figure 25.4   Correct for speed and heavy vehicle content. Correction = 4·7 dB(A).
From figure 25.5   Correct for distance over hard ground (assume height of reception point is 2 m). Correction = −5·5 dB(A).

Reflection effect at facade = +2·5 dB(A).
L_{10} (18 h) at facade due to stream 1 = 71·6 dB(A).
Similarly for stream 2.

From figure 25.3 L_{10} (18 h) = 72·9 dB(A).
From figure 25.4   Correction = 0·5 dB(A).
From figure 25.5   Correction = −6·0 dB(A).

Reflection effect at facade = +2·5 dB(A).
L_{10} (18 h) at facade due to stream 2 = 69·9 dB(A).
From figure 25.8   Combined L_{10} (18 h) at facade = 74 dB(A).

7. The improvement in noise exposure that will be obtained by the use of a very long barrier adjacent to a motorway is calculated from the geometrical layout shown in figure 25.13.

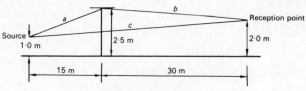

Figure 25.13

The improvement is related to $(a + b) - c$ and

$$a^2 = 15^2 + (2 \cdot 5 - 1 \cdot 0)^2$$
$$= 225 + 2 \cdot 25$$
$$= 227 \cdot 25$$
$$\therefore a = 15 \cdot 074 \text{ m}$$
$$b^2 = 30^2 + (2 \cdot 5 - 2 \cdot 0)^2$$
$$= 900 + 0 \cdot 25$$
$$\therefore b = 30 \cdot 004 \text{ m}$$
$$c^2 = 45^2 + (2 \cdot 0 - 1 \cdot 0)^2$$
$$= 2025 + 1 \cdot 0$$
$$= 2026$$
$$\therefore c = 45 \cdot 011 \text{ m}$$
$$(a + b) - c = 0 \cdot 067 \text{ m}$$

From figure 25.7 it can be seen that there is a reduction in the L_{10} noise exposure of approximately $9 \cdot 5$ dB(A).

8. The noise attenuation due to an unsymmetrical barrier may be calculated from a consideration of the relative geometric layout of the barrier and reception point.

$$\tan A_1 = \frac{200}{30} = 6 \cdot 6667$$
$$A_1 = 81° \, 28'$$
$$\tan A_2 = \frac{50}{30} = 1 \cdot 6667$$
$$A_2 = 59° \, 0'$$
$$\theta_B = 140° \, 28'$$

Assume hard ground conditions, then $\theta_S = 0°$, $\theta_H = 39° \, 32'$.
Let L_{10} be the noise level at the reception point when the barrier is not present. Correction for angle of view is

$$L_U = L_{10} + 10 \log (39°32'/180°) = L_{10} - 6 \cdot 6 \text{dB(A)}$$

Unobstructed noise level at the reception point corrected for barrier and field of view attenuation L_B is: $L_{10} - 9 \cdot 5 + 10 \log (140°28'/180°) = L_{10} - 10 \cdot 6$ dB(A).
Combining L_U and L_B gives shielded L_{10}. Then attenuation due to barrier is given by: shielded $L_{10} - L_{10} = (L_{10} - 10 \cdot 6) + (L_{10} - 6 \cdot 6)$

$$= 10 \log (10^{-1 \cdot 06} + 10^{-0 \cdot 66})$$

or, the attenuation of the barrier is $5 \cdot 1$ dB(A).

26

The environmental effects of highway traffic pollution

Air pollution from road traffic

During recent years there has been a widespread attempt to reduce air pollution from all sources. In the United Kingdom the Clean Air Act of 1956 has resulted in a noticeable decrease in coal consumption and a reduction in air pollution from domestic and industrial sources. During this same period there has been a marked increase in the volume of road traffic and consequently an increase in pollution from this source. The National Society for Clean Air[1] estimate that, during 1965 in the United Kingdom, the total coal and oil consumed was equivalent to 276 million tons (coal equivalent) and only 26 million tons of this were used for road or rail transport. It is however an increasing source of pollution, which is emitted in situations close to human activity. Approximately one-third of the carbon monoxide in the atmosphere is produced from vehicle exhausts.

The major sources of atmospheric pollution caused by motor vehicles have been given by Sherwood and Bowers[2] and may be classified as:

a) exhaust gases;
b) evaporative losses from the fuel tank and carburettor;
c) crank case losses;
d) dust produced by the wearing away of tyres, brake linings and clutch plates.

Considering the exhaust gases, the following compounds are normally present in the discharge from vehicle exhausts:

a) carbon dioxide;
b) water vapour;
c) unburnt petrol;
d) organic compounds produced from petrol;
e) carbon monoxide;
f) oxides of nitrogen;
g) lead compounds;
h) carbon particles in the form of smoke.

On occasions these components of the exhaust may react with each other to produce unpleasant secondary products. The most well-known effect of this type is the Los Angeles 'smog', which, because of the bright sunlight and the topography of the region is formed by the reaction of the oxides of nitrogen and some of the hydrocarbons.

Both petrol and diesel engines give rise to similar products in their exhausts but the relative proportions differ. Diesel engine exhaust gases contain significantly lower proportions of pollutants than do those produced by petrol engines. Unfortunately, an incorrectly operated or maintained diesel engine is liable to emit smoke and produce an offensive smell but even then, apart from carbon particles, the degree of pollution is less than that produced by petrol engines.

The effects of these pollutants have been reviewed at the Transport and Road Research Laboratory[2] and the conclusions will be summarised.

Unburnt fuel and secondary products produced from the fuel

Unburnt fuel is emitted to the atmosphere by evaporation from the fuel tank and carburettor. A high proportion of the hydrocarbons in the crank case blow-by and in the exhaust gases also consists of unburnt fuel. The constituents of petrol are not considered to be toxic, but some of them have slight anaesthetic effects in high concentrations.

Many compounds are found in the gaseous products of the fuel emitted in the exhaust gases. Of these a significant proportion of aldehydes are produced. These have an irritant action on the eyes and on the respiratory system and they can be smelt in very small proportions.

In addition to the gaseous products a number of polynuclear aromatic compounds also emitted with the exhaust gas in the form of very fine particles, which can persist in the air for lengthy periods. They are important because some of them, such as benzpyrene, are known to be carcinogenic, but the extent of the health hazard for the proportions present is not known. Campbell and Clemmesen[3] have also shown that the benzpyrene content in the air of road tunnels is much less than the content in the air in industrial areas.

Carbon monoxide

The dangers of the absorption of carbon monoxide and its reaction with haemoglobin in the blood are well known. The degree of absorption depends on the carbon monoxide content in the air, the period of exposure and the activity of the individual. A survey the Transport and Road Research Laboratory of the carbon monoxide content of the air in busy city streets in the United Kingdom has indicated that at the present levels, road users will not be aware of any discomfort from this source, but this may not be true for policemen and others operating in city streets for long periods of time.

While it is believed that carbon monoxide is unlikely to leave any permanent effect or cause acute physical discomfort, its effect cannot be entirely discounted because relatively small concentrations of carboxy-haemoglobin in the blood have been shown to temporarily impair mental ability. Fortunately this is only likely to occur in still weather in traffic jams, and even then only when the subject has been working hard for an hour.

Carbon monoxide might be a greater cause for concern if its level were increasing, but it quickly dissipates. Sherwood and Bowers[2] conclude that carbon monoxide produced by motor vehicles is unlikely to be a medical danger unless unsuspected synergistic effects are present. It may however be socially objectionable because of its effect on a small proportion of particularly susceptible members of the public.

Oxides of nitrogen

Both nitric oxide and nitrogen dioxide are produced by the internal combustion engine, the former in much greater quantities but nitric oxide oxidises to nitrogen dioxide. Typically in a city street there is twice as much nitric oxide as there is nitrogen dioxide.

Nitrogen dioxide is considerably more toxic than nitric oxide, but the concentration of the former in city streets is less than 1 per cent of the maximum allowable concentration for an industrial 8-hour exposure.

Little information is available on the effect of oxides of nitrogen on health—the concentrations are very low—but the effect is clearly adverse and the long-term effects of exposure should be clearly kept in mind.

Lead compounds

Vehicle exhaust gases[4] are the major source of lead in urban areas, originating from the 'anti-knock' components of the fuel. About 2 g of the metal is present in each gallon of petrol and from 25 to 50 per cent of this is discharged to the atmosphere in the form of a fairly stable aerosol of lead halides and oxide. It has been stated[2] that the concentration of lead in the air of city streets is approximately 2–4 ng/m^3 and this should be compared with the maximum allowable concentration for three hours daily exposure of 300 ng/m^3. The level that exists in city streets however is some 20 times greater than the background level in rural areas, but it is argued that whether or not these levels are dangerous as such, this level of increase over natural levels is likely to result in the concentration of lead in some biological chains leading to toxic doses in some edible foods.

Smoke

The incomplete combustion of fuel results in the emission of very fine particles of carbon. It is a problem that is usually associated with diesel engines and it is linked with other forms of pollution by the general public. In itself the smoke is not considered a hazard to health but the carbon particles act as nuclei for haze formation and adsorb sulphur dioxide and nitrogen oxides, which may cause lung damage. Little is known of this phenomenon and Sherwood and Bowers[2] considered it may be a more important medical problem than previously has been suspected.

Other particulate matter

In addition to finely divided carbon and unburnt fuel in the exhaust emission, road traffic produces other particulate matter. This takes the form of finely divided rubber from the vehicle tyres and also asbestos dust from brake linings and clutch plates.

Prolonged exposure to asbestos dust is known to be harmful to health but the proportions produced by road traffic are very small and there is no evidence that it is a hazai
to health.

Prevention of exhaust pollution

Because of the acute problems caused by vehicle exhaust pollution in California this
State has pioneered the reduction of exhaust pollution by legal restriction. Restrictioi
in the remainder of the United States are being progressively applied and it is estimate
that by 1985 the levels of pollution will reach a minimum before rising again because
of an increase in vehicles.

It appears likely that many other countries will limit exhaust pollution even thoug
the benefits of such action are not easily defined. Pollution from individual engines is
likely to be reduced by the modification of existing engines or the development of ne
engine types. One source of pollution however, lead, can be eliminated by omitting it
from petrol and maintaining the same octane rating by more expensive means. Altern
tively engines with lower compression ratios can be used.

More general measures that can be used to reduce exhaust pollution include: the u
of smaller engines and vehicles in congested urban areas; the use of electrically driven
vehicles; the improvement of vehicle flow; restrictions on the use of private vehicles i
the central areas of cities.

In conclusion it can be stated that the general consensus of opinion among those
authorities who should be able to form an accurate opinion is that air pollution deriv
from the internal combustion engine is not a serious health hazard in the United
Kingdom. The reason for this conclusion is that no one has shown that any of the
pollutants present in vehicle exhausts have harmful permanent effects in the proporti
in which they are found at present. Exhaust fumes have, however, adverse social effe
and at the moment it appears that this will be the main reason for the introduction o
legislative controls on pollution from highway vehicles.

References

1. National Society for Clean Air. Air pollution from road vehicles—a report by the
   technical committee of the National Society for Clean Air. London (1967)
2. P. T. Sherwood and P. H. Bowers. Air pollution from road traffic—a review of the
   present position. *Rd Res. Laboratory Report* 325 (1970)
3. J. M. Campbell and J. Clemmesen. Benzpyrene and other polycyclic hydrocarbon
   the air of Copenhagen. *Dan. med. Bull.* (November 1956), 205–11
4. American Chemical Society. Cleaning our environment. The chemical basis for ac
   Report by the sub-committee on Environmental Improvement, Committee on
   Chemistry and Public Affairs, American Chemical Society, Washington (1969)

Problems

Are the following statements correct or incorrect?

1. The major source of pollution adjacent to heavily trafficked motorways is the
diesel-engined goods vehicle.

2. Traffic 'smog' is likely to occur in regions where vehicle mileage is considerable and there is a low incidence of sunlight.

3. The benzpyrene content of the air in highway tunnels gives cause for alarm because it is greater than is found in industrial areas.

4. The carbon monoxide produced by vehicle exhausts in busy city streets is not likely to cause ill effects to vehicle drivers under free-flowing traffic conditions.

5. The concentration of oxides of nitrogen produced by vehicle exhausts as found adjacent to the highway is less than 1 per cent of the maximum allowable concentration for an 8-hour exposure in industrial conditions and has no long term effects.

6. The discharge from vehicle exhausts contains a stable aerosol of lead halides, which gives cause for concern because it produces a concentration of lead in the air of city streets which is considerably greater than the average level in rural areas.

7. It is the opinion of health authorities that in the United Kingdom pollution from vehicle exhausts is not a serious health hazard and for this reason it can be expected that legislation will generally be required to limit pollution from this source on purely social grounds.

Solutions

1. This statement is incorrect. Diesel-engine exhaust gases contain significantly lower proportions of pollutants than those produced by petrol engines.

2. This statement is incorrect. Traffic 'smog' is caused by the reaction of oxides of nitrogen and some of the hydrocarbons in the presence of bright sunlight.

3. This statement is incorrect. The benzpyrene content of the air in highway tunnels is less than that found in industrial areas.

4. This statement is correct. A survey carried out by the Transport and Road Research Laboratory has indicated that the carbon monoxide content of air in busy city streets in the United Kingdom is not likely to cause discomfort to road users when traffic is flowing freely.

5. This statement is incorrect. While the concentration of oxides of nitrogen produced by vehicle exhausts is as stated, the long-term effects are unknown.

6. This statement is correct.

7. This statement is correct.

27

Traffic congestion and restraint

The motor car is an invention which, within half a century, has revolutionised our way of life. It has made possible a dispersal of dwellings far exceeding that of the railway age, it has offered a wide choice of employment situations and has increased the scope of recreational activities to a remarkable extent.

At the same time the growth of vehicle ownership and use, together with population increases and the attraction of human activity into urban regions has resulted in considerable problems. The polluting effect of vehicle exhausts, the noise associated with road vehicles and above all the demand for physical space, all make it necessary to control the use of vehicles in urban areas.

The demand for road space, especially in existing central town areas, will always be greater than the supply because even if the necessary financial resources were available there would be conflicting demands for the available land. The fact that demand for road space is greater than the supply results in traffic congestion. While traffic management and urban highway construction have their place in minimising congestion it is now generally accepted that, without the dispersal of town centre activities, the only solution at the present time is a greater emphasis on public transpor

If this transfer from individual to public transport is accepted as one part of a solution to the problems of traffic in towns, then it will be necessary to find some means o traffic restraint. At the present time congestion itself acts as a restraint, causing trips which would take place at congested periods to be made at other times, or by alternati non-congested modes, or the trips may not be made at all. Congestion is however an inefficient form of restraint in that the priority of service is first come, first served, regardless of the value of the trip to either the tripmaker or the community. It is inefficient in the use of resources and is detrimental to the environment adjacent to the facility.

There are three general ways in which restraint could be applied. Firstly the entry o vehicles to certain areas at certain times could be prohibited by administrative means. On a limited scale this is already frequently employed in the form of pedestrian precinc but its application on a wide scale would involve the entry of specialist service or emergency vehicles and would involve decisions as to who should be allowed entry on the grounds of the value of their trip either to themselves or to the community.

Secondly restraint could be applied by the use of parking regulations. The Ministry of Transport report, 'Better Use of Town Roads'[1], considered that the most promising method of restraint, at least for the shorter term, would be to intensify control over th location, amount and use of parking space, both on and off the street. They especially considered it necessary to restrict long-term parking, which is characteristic of car commuting. They felt that in some places control over the use of publicly available parking space might not be adequate and might have to extend to privately available parking space, even though this would be costly and require new legislation.

The third form of restraint that could be applied is road pricing. This is a form of road user taxation whereby users of congested roads would be charged according to the distance travelled or the time spent on them, at varying rates governed by the degree of congestion. A system of road pricing does however exist; it was introduced by Lloyd George in 1909 and basic features of the system are still unchanged. It is not however an efficient form of road pricing because the annual licence is not related to road use and petrol tax does not discriminate to any significant extent between tripmaking in congested and non-congested conditions. It has been said[2] that, 'the issue is not whether we should have a system of road pricing, because we already have one, but whether we could devise a better one'.

Road pricing is an attempt to change the principle that highways should be treated as welfare services, that is financed out of taxes, to the principle that they should be treated as public utility services for which charges are made. It has been argued[3] that road pricing is democratic because it is the tripmaker who makes the decision as to whether or not the trip should be made at the given price rather than a government body making a decision that his trip was in the interests of the community.

If the price is placed at a level that reflects the costs a tripmaker imposes on others, it will produce traffic flows that reflect social benefits and costs as well as the trip-maker's private benefits and costs. If alternative means of transport are also priced in this way then the resultant traffic flows would give a better indication of the need for the construction of future transport facilities. This is because the true demand for transport facilities will be known rather than relying on projections of 'free' tripmaking in which the tripmaker does not bear any of the cost imposed on others.

Some doubts on road pricing have been stated by Lichfield[4]. These are that conditions approximating to perfect competition are difficult to visualise for transportation systems. It is also difficult to estimate a price charge that reflects the indirect repercussions caused to external economies or diseconomies in production or consumption. When environmental costs have to be included as well as congestion costs then determination of the pricing charge requires further research. Finally income distribution is likely to be affected when a uniform pricing charge is imposed upon all tripmakers.

The Ministry of Transport report, 'Road Pricing: The Economic and Technical Possibilities'[5], gives the following requirements for a pricing system.

1. Charges should be closely related to the amount of use made of the roads.
2. It should be possible to vary the price for different roads, at different times of day/week/year or for different vehicle classes.
3. Prices should be stable and ascertainable by road users before they commence the journey.
4. Payment in advance should be possible although credit facilities should in certain cases be permissible.
5. It should be accepted as fair.
6. It should be simple.
7. Equipment used should be reliable.

8. The system should be reasonably free from fraud and evasion, both deliberate and unintentional.

9. If necessary it should be capable of being applied to the whole country.

They also considered that the following would be desirable, but not so important, requirements.

1. Payment should be in small amounts, at fairly frequent intervals, say £5/month (19◄ prices).
2. Drivers in high-cost areas should be aware of the amount they are paying.
3. Drivers' attention should not be diverted by the pricing system.
4. The system should be applicable to drivers from abroad.
5. Enforcement should lie with traffic wardens.
6. If possible it would be preferable to use the method for charging for street parking.
7. It should give a measure of the strength of demand for road space, so as to guide the planning of new routes.
8. The system should be capable of gradual introduction starting with an experimental stage.

In addition to parking charges the Ministry Group examined the possibility of entry charges using a system of supplementary licences, its main advantage over parking charges being that it restrains both non-parking tripmakers and those who are able to park privately. Its disadvantage is that it fails to take into account different amounts of road space within the licensing area, or use at different levels of congestion. The first disadvantage limits the size of the licensing area, and in addition the larger the area, the higher will be the licence fee and the greater the effects at the boundaries. In 'Better Use of Town Roads' it is suggested that the maximum practicable area for such a system might be no more than 10 square miles.

Another indirect method of charging for road use which was examined was the employment of differential fuel taxes. A different petrol tax could be levied in different areas according to the congestion in the area. This tax however could be avoided by filling up in low-tax areas or it could be avoided by special fuel-carrying journeys unless the differential was small. This is only a general tax on a large area, including congested and uncongested roads in the same area at congested and uncongested times. The Report considered that it would affect individual garages, cause a black market in petrol and encourage people to carry cans of petrol. In addition commuters would find it easy to avoid tax unless the tax area was very large.

The Panel considered that electronic metering offered the best opportunity for meeting the requirements for a pricing system. They outlined seven systems, five of which were automatic and two were driver operated. A distinction was made between continuous systems in which the charge is made on the basis of time spent or distance travelled in the charging area, and point systems in which a driver is charged every time he passes a pricing point.

In subsequent development work the Transport and Road Research Laboratory[6] rejected driver-operated and continuous automatic systems. Their reasons for this decision were given as:

(a) Driver-operated systems impose an additional demand on the driver and may distract his attention. In addition they are vulnerable to attempts at evasion.

(b) Systems based on measuring time or distance require time references or distance-measuring methods to be used in the vehicle.

(c) There is the difficulty that a charge should not be registered when the vehicle enters private premises: this is a particularly difficult problem when a time-based system is used.

(d) The time-based system would result in an attempt to minimise time with a consequent decrease in safety.

(e) Continuous systems are not fail-safe in that failure to record the departure of a vehicle from the charging zone would result in excessive charges being levied.

These disadvantages may be avoided by the use of a point-pricing system, whose advantages are:

(a) Point-pricing systems are flexible because the charging rate can be varied by alteration in the spacing of the pricing points. The system can also be easily extended by the addition of further pricing points.

(b) Variations in the charging rate can be obtained by varying the price per charging point or by switching out series of points at non-congested times of the day.

(c) Relatively simple apparatus is required on the vehicle for this type of system.

(d) The point-pricing system is considered to be consistent with existing road charging facilities and could be an extension of a system of supplementary licences. Regular entrants to the pricing zone would begin to mount special equipment in their vehicles while infrequent entrants would purchase occasional daily licences. In some circumstances it would be possible to use the point-pricing system to levy parking charges.

(e) A point-pricing system is fail-safe in that if a vehicle fails to register passing a given point, the tripmaker gains.

The point-pricing system makes it possible to place pricing points anywhere on the road network but they should be placed so as not to result in undesirable route selection, preferably in the form of closed cordons. This would result in the charging area being divided into small zones with a pricing point imposing a charge every time a vehicle passes from one zone to another.

At the Transport and Road Research Laboratory two point-pricing systems are being developed. One is an on-vehicle system in which the meter that stores the number of pricing points passed is mounted on the vehicle. The other is an off-vehicle system in which each vehicle that passes a pricing point is uniquely detected and the information is then passed on to a central computer.

While the implementation of a full road-pricing scheme is obviously still far distant the report 'Better Use of Town Roads' considered direct pricing as potentially the most efficient means of restraint.

If any road-pricing system is to be introduced it will require considerable capital investment and incur substantial maintenance costs. It must also be shown that the net economic benefits of the scheme, that is the economic benefits minus the operating costs, are worth while when compared with other forms of restraint.

The variations of net benefit that occur with increasing traffic flow on a highway can be illustrated by the following example. Suppose that all tripmakers consider the value of a trip on this highway is 50 p. The tripmakers' private costs of the journey are normally composed of two elements. Firstly there are vehicle operating costs including fuel (but excluding tax), maintenance and depreciation which may be expressed as a cost per vehicle kilometre. The second element consists of those costs that vary with journey speed, and mainly depend on the cost of individuals' travel time. For this example the travel costs given in 'The Economic Appraisal of Inter-Urban Road Improvement Schemes'[7] will be used; they are reproduced as table 27.1. These travel costs comprise both the fixed and variable private costs.

It will also be assumed that the relationship between travel speed and flow for this highway has the form

$$v = 87 - \frac{q + 1400}{82} \quad \text{or} \quad 75 \text{ km/h} \tag{27.1}$$

whichever is the least, where q is the flow in vehicles/hour.

The variation of costs and benefits as the flow along this highway link increases are calculated and tabulated in table 27.2.

TABLE 27.1   Private travel costs at varying speeds

Vehicle speed (km/h)	Travel costs (p/km)		
17	6·78	54	2·88
18	6·45	55	2·85
19	6·16	56	2·83
20	5·91	57	2·80
21	5·68	58	2·78
22	5·47	59	2·75
23	5·28	60	2·73
24	5·11	61	2·70
25	4·96	62	2·68
26	4·82	63	2·65
27	4·69	64	2·63
28	4·56	65	2·60
29	4·44	66	2·58
30	4·32	67	2·56
31	4·21	68	2·54
32	4·11	69	2·52
33	4·01	70	2·51
34	3·93	71	2·49
35	3·85	72	2·47
36	3·78	73	2·46
37	3·70	74	2·44
38	3·63	75	2·42
39	3·56	76	2·41
40	3·50	77	2·39
41	3·44	78	2·37
42	3·38	79	2·36
43	3·32	80	2·34
44	3·26	81	2·33
45	3·20	82	2·31
46	3·15	83	2·30
47	3·10	84	2·29
48	3·06	85	2·28
49	3·02	86	2·27
50	2·99	87	2·26
51	2·96	88	2·25
52	2·93	89	2·25
53	2·90	90	2·24

TABLE 27.2 Variations of gross and net benefits with increasing flow

1	2	3	4	5	6	7	8
						Increase due to the addition of one vehicle	
Flow q (veh/h)	Speed–equation (27.1) (km/h)	Cost/km vehicle (p)	Total cost (1) x (3) x 5 km (£)	Gross benefit 50 p x (1) (£)	Net benefit (5) – (4) (£)	Cost (£)	Net benefit (£)
100	69	2·52	12·60	50	37·40		
200	67	2·56	25·60	100	74·40	0·13	0·37
500	64	2·63	67·75	250	182·25		
600	63	2·65	79·50	300	220·50	0·14	0·36
1000	58	2·78	139·00	500	361·00		
1500	52	2·93	219·75	750	530·25		
1600	50	2·99	239·20	800	560·80	0·19	0·31
2000	46	3·15	315·00	1000	685·00		
2500	39	3·56	445·00	1250	805·00		
2600	38	3·63	471·90	1300	828·10	0·27	0·23
3000	33	4·01	601·50	1500	898·50		
3100	32	4·11	637·05	1550	912·95	0·36	0·14
3500	27	4·69	820·75	1750	929·25		
3600	26	4·82	867·60	1800	932·40	0·47	0·03
3700	25	4·96	917·60	1850	932·40	0·50	0·00
4000	21	5·68	1136·00	2000	864·00		
4100	20	5·91	1211·55	2050	838·45	0·76	−0·26
4300	17	6·78	1457·70	2150	692·30		

It can be seen from table 27.2 and figure 27.1 that the net benefits from operating the system increase as the traffic flow increases until an optimum flow value, from the point of view of net benefits obtained, of approximately 3700 veh/h is reached.

The additional cost of adding one extra vehicle to the traffic stream is referred to as the marginal cost and in this example it is calculated by noting the change in costs and

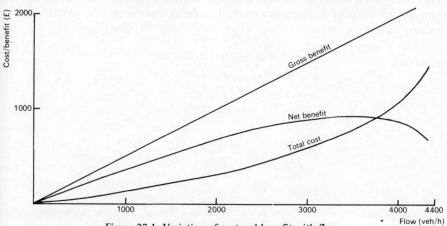

Figure 27.1 Variation of cost and benefit with flow

benefits when the flow increases by 100 veh/h. It should be noted that since the rate of change of costs is not uniform this will only give an approximate value.

From table 27.2 it can be seen that at low levels of flow the marginal or additional cost of one vehicle is considerably less than the benefit of the journey. At flows exceeding 3700 veh/h the marginal cost is greater than the benefit of the journey, and additional vehicles impose greater costs than the benefits they receive. These journeys would still be made however, because even at a flow of 4300 veh/h the cost/km to each driver is 6.78 p and the total journey cost is 34.90 p, which is less than the benefit of 50 p he would receive by making the trip.

In a real situation however not all tripmakers would attach the same value to the benefit they would obtain from the trip. Because of differences in income and in the nature of the trip as well as differences in the value of getting to the destination the demand for tripmaking may be expected to fall as the cost of the trip increases.

The demand D for tripmaking may be expressed as a function of the private cost or price p of the trip, that is

$$D = f(p) = Kp^{-\gamma}$$

where K and γ are parameters. The general form of this relationship is given in figure 27.2.

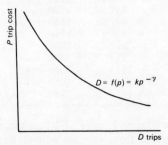

Figure 27.2 General form of the demand function

When the effects of varying trip costs are being considered an important factor in determining the relationship between a change in trip cost and the resulting change in tripmaking is the price or cost elasticity e_p. It is defined as the percentage change in quantity demanded that results from a 1 per cent change in price. Then

$$e_p = \frac{\partial D/D}{\partial p/p}$$

$$= -\frac{p}{D} \times \gamma Kp^{-\gamma-1}$$

$$= -\frac{K\gamma p^{-\gamma}}{D}$$

$$= -\gamma$$

If the demand function is K/p, as is frequently assumed, then the price elasticity is unity.

The cost to the tripmaker has been referred to as the private cost of a journey while the cost that a tripmaker imposes upon other tripmakers because of an increase in congestion is referred to as the congestion cost.

As has been indicated previously the addition to the total costs caused by one extra tripmaker is the marginal cost. It consists of the private costs of the additional trip together with the congestion costs caused by the additional trip.

In addition to these costs there are environmental costs that the trip imposes upon the area adjacent to the highway, and road maintenance costs.

Private costs are divisible into two parts, those proportional to distance, such as fuel, maintenance, depreciation, and those varying with journey speed, which are chiefly associated with the value of time.

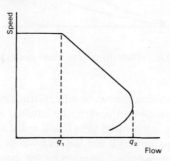

Figure 27.3 A general form of the speed/flow curve

Congestion costs arise from the interaction of vehicles and they are illustrated in the typical relationship between speed and flow shown in figure 27.3, where q_1 is the flow at which interaction commences. It may be assumed that there is a linear relationship between q_1 and q_2, the value of maximum flow. This relationship may be assumed to be of the form

$$v = a - bq \qquad (27.2)$$

If it is assumed that the total private cost of a trip may be expressed as

$$c + \frac{d}{v} \text{ pence/km} \qquad (27.3)$$

where c is the component that is proportional to distance,
    d is the component that is proportional to time,

then the private costs of a tripmaker are

$$c + \frac{d}{a - bq} \text{ pence/km}$$

so that the private cost/flow curve has the form shown in figure 27.4. The intersection of the demand curve and the private cost/flow curve at E represents the equilibrium condition at cost c_e and flow q_e. Additional trips above flow q_e will not be made because the private cost of the trip is greater than the benefit of the trip as given by the demand curve.

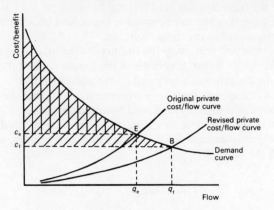

Figure 27.4 The change in net benefit to tripmakers due to a highway improvement

At this equilibrium condition the net benefit to tripmakers, that is the benefits minus the private costs, is represented by the vertically hatched area.

If however the highway is improved then it will result in a revised private cost/flow curve with a new equilibrium point B and the flow will be q_f at a trip cost c_f. In this case the net benefit to tripmakers will be given by the diagonally hatched area.

The marginal cost/flow curve may be obtained by differentiating the expression for the total cost of tripmaking

$$\left(c + \frac{d}{a - bq}\right) q$$

with respect to q, giving the equation of the marginal cost/flow curve

$$c + \frac{d}{v} \times \frac{a}{v} \qquad (27.4)$$

It can be seen that this is of the same form as the cost/flow curve with the private cost component multiplied by the factor a/v. From a knowledge of the speed/flow relationship a/v will be greater than unity and the marginal cost/flow curve will have the general relationship to the private cost/flow curve shown in figure 27.5.

In figure 27.5 the equilibrium flow and private cost is given by the intersection of the marginal cost/flow and demand curves at E, a situation which would occur when a decision as to whether a trip should be made is based solely on private cost. The intersection of the marginal cost/flow curve with the demand curve at O however gives the optimum flow. If flow is restricted to below this level then only trips with a marginal cost less than the value of the trip to the tripmaker will be allowed. If the flow is greater than the optimum then a trip is allowed that has marginal costs greater than the value of the trip.

If traffic flow conditions are shifted from the equilibrium position E to the optimum position O then the change in net benefit will be represented by the area A-c_p-P-O

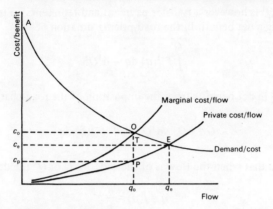

Figure 27.5 The relationships between marginal and private costs and flow, and between demand and cost.

minus the area A-c_e-E. This is in fact equal to the area c_e-c_p-P-T minus the area OTE, which represents the gain in net benefit enjoyed by the tripmakers who are not removed due to improved speed, minus the net benefit previously enjoyed by the trip-makers who are now removed.

The movement from E to O can be achieved by the use of a road pricing charge, represented by PO. Individual tripmakers will however be worse off under the pricing system because the decrease in private costs TP will always be less than the pricing charge OP.

An estimate of the economic benefits to be derived from direct road pricing is contained in 'Road Pricing: The Economic and Technical Possibilities'[5] and the following illustration of a method which may be used to calculate the net benefits from the imposition of a road-pricing charge is based on that report.

A highway network is considered where the flow is q and the corresponding private costs of a tripmaker are p. The demand curve is of the form $q = f(p)$, which may alternatively be expressed as the inverse relationship $p = f^{-1}(q)$, where $f^{-1}f(p) = p$.

Consider the situation before any price is imposed and let the private cost be α and the flow be Q. From the reasoning given previously, the net benefit to all tripmakers is

$$\int_0^Q f^{-1}(q)\, dq - \alpha Q \tag{27.5}$$

Consider now the situation in which a road pricing charge per vehicle/km of β is imposed, causing a decrease in flow to Q^1 and an increase in speed, and a reduction in costs from α to α^1 (let $\alpha - \alpha^1 = g$). As before the net benefit to tripmakers is now

$$\int_0^{Q^1} f^{-1}(q)\, dq - (\alpha^1 + \beta)Q^1$$

The amount βQ^1 is however a transfer payment and represents no real cost to the community, giving a net benefit in the road-pricing situation of

$$\int_0^{Q^1} f^{-1}(q)\, dq - \alpha^1 Q^1 \tag{27.6}$$

The increase G in net benefits from the imposition of the road charge β is then

$$G = \alpha Q - \alpha^1 Q^1 - \int_{Q^1}^{Q} f^{-1}(q)\, dq \tag{27.7}$$

If it is assumed that when the flow is of the order of Q or Q^1 the demand function is

$$q = f(p) = \frac{k}{p}$$

then the elasticity of demand is unity and

$$Q = \frac{k}{\alpha}, \qquad Q^1 = \frac{k}{\alpha^1 + \beta}, \qquad p = f^{-1}(q) = \frac{k}{q}$$

By substituting in equation 27.7 it can be shown that the increase in net benefits is

$$G = \beta Q^1 + \alpha Q \log \frac{Q^1}{Q} \tag{27.8}$$

and therefore

$$\frac{G}{Q} = \beta \frac{Q^1}{Q} + \alpha \log \frac{Q^1}{Q} \tag{27.9}$$

where Q^1/Q is the ratio of the new flow to the old flow.

In order to quantify the above relationship a process of inspection is necessary because Q^1/Q varies with β. In the first place the relationship between g and Q^1/Q must be established from speed/cost and speed/flow relationships. The report took a speed/cost relationship derived by Charlesworth and Paisley[8] and updated it to values at the date of the report (1964) to give

$$\alpha = \frac{1}{13 \cdot 92}\left(4 \cdot 1 + \frac{224}{v}\right) \tag{27.10}$$

where α was the average private cost in shillings for one passenger car unit travelling at v miles/h. The speed/flow relationship employed was one derived by Thomson[9] for Central London

$$v = 28 - \frac{q}{125} \tag{27.11}$$

where q is the flow in p.c.u.s per hour and v is the speed in miles/h.

It was then possible to calculate the average private cost (from equation 27.10) and the corresponding flow (from equation 27.11) for a range of speeds. For the difference between any two journey speeds of which the lower is the congested or base speed the difference in the private costs (g) and the ratio of the two flows (Q^1/Q) may be tabulated.

For varying base or congested speeds it is then possible to insert values of Q^1/Q, corresponding values of α and varying values of β into equation 27.9 to obtain G/Q, the net benefits per vehicle mile. The variation of net benefits for differing pricing charges β derived[5] on this basis is illustrated in figure 27.6.

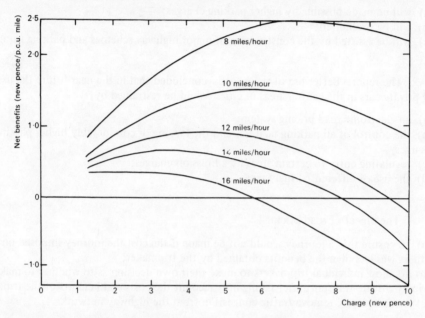

Figure 27.6 Net benefits from different charges at different speeds (adapted from ref. 5)

References

1. Ministry of Transport. Better use of town roads. H.M.S.O. (1967)
2. J. M. Thomson. Case for road pricing. *Traff. Engng Control* (March 1968), 536–9
3. G. J. Roth. *Paying for Roads: The Economics of Traffic Congestion*. Penguin, Harmondsworth (1967)
4. N. Lichfield. Planner/economists view of road pricing. *Traff. Engng Control* (February 1968), 485–7
5. Ministry of Transport. Road Pricing: The Economic and Technical Possibilities. H.M.S.O. (1967)
6. G. Maycock. Implementation of traffic restraint. *Report* LR 422, Transport and Road Research Laboratory (1972)
7. Department of the Environment. The Economic Appraisal of Inter-Urban Road Improvement Schemes (1971)
8. G. Charlesworth and J. L. Paisley. The economic assessment of returns from road works. *Proc. Instn civ. Engrs.*, **14** (1959), 229–54
9. J. M. Thomson. Calculations of economic advantages arising from a system of road pricing. Paper PRP. 18, 1962. The Transport and Road Research Laboratory

Problems

Select the correct completions to the following statements.

1. In the central areas of existing cities the future use of the private motor vehicle:

(a) will be made possible by higher parking charges;
(b) will be prohibited;
(c) will be assisted by the construction of major highway schemes and parking garages.

2. The report 'Better use of town roads' concluded that in the near future the use of private cars in the central areas of cities would be restrained by:

(a) an electronic road pricing system;
(b) the control of all parking facilities and the levying of considerably higher parking charges;
(c) regulating entry to certain areas by admission charges;
(d) the use of differential fuel taxes.

3. The object of road pricing is:

(a) to ensure that a journey would not be made if the cost the journey imposed on others was less than the benefits obtained by the tripmaker;
(b) to allow individual tripmakers to make their own decision as to whether to make a trip by private transport after taking into account the costs and benefits of the trip;
(c) to completely remove traffic congestion from the highway network.

4. A speed/flow relationship for a highway has the form shown in figure 27.7.

When the flow on the highway is:

(i) in the region 0–q_1
(ii) in the region q_1–q_2

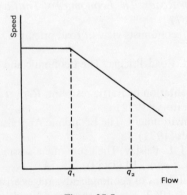

Figure 27.7

will the additional cost of introducing an extra vehicle into the flow be:

(a) the private costs of the additional vehicle?
(b) the private costs of the additional vehicle together with an addition to the private costs of other vehicles?
(c) negligible?

5. Indicate on figure 27.8 which curve is:

(a) a demand curve on which tripmakers all value the benefits of their trip equally;
(b) a demand curve on which some tripmakers value the benefits of their trip at a higher level than others;
(c) a cost/flow curve on which costs are a function of journey distance and time;
(d) the marginal cost/flow curve for the above cost/flow curve.

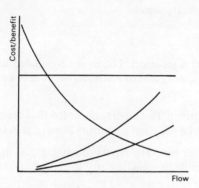

Figure 27.8

6. A highway link has the following characteristics:

(a) a speed/flow relationship of $v = 50 - q/100$
where v is the stream journey speed (km/h)
q is the stream flow (p.c.u./h)
(b) a flow/cost relationship of $p = 1 \cdot 25(5 + 300/v)$
where p is the private cost in p.c.u./km
v is the journey speed in km/h.
(c) a demand relationship of the form

$$D = 4 \times 10^4 p^{-1}$$

where D is the demand in p.c.u./h
p is the private cost in km/h.

Determine graphically:

(a) the flow and the speed on the highway when each tripmaker is only aware of his private costs;
(b) the flow and the speed on the highway when a trip is only made if the benefit of a trip to the tripmaker exceeds the additional cost imposed on other tripmakers;
(c) the pricing charge that would result in the flow (b).

Solutions

1. (a) This statement is correct. The report 'Better Use of Town Roads' concluded that for the short term the most promising form of restraint is likely to be the intensification of controls over parking. It is this restraint over the less valued trips that will make it possible to use private motor vehicles for the more valued trips.

(b) This statement is incorrect. It would not be practicable to prohibit the entry of all private motor vehicles into larger town areas and entry by permit would raise administrative difficulties.

(c) This statement is incorrect. Large-scale highway and parking garage construction would be necessary on so large a scale that it is doubtful if the financial resources could be made available. If it were financially possible then it would result in the wholesale reconstruction of central city areas.

2. (a) This statement is incorrect. The report concluded that this method is potentially the most efficient form of restraint, but that it would take several years to develop.

(b) This statement is correct. The report concluded that the strict control of parking would in the short term be the most promising form of restraint.

(c) This statement is incorrect. The report concluded that they were far from satisfied that it would be practicable to regulate entry to certain areas by admission charges.

(d) This statement is incorrect. The report did not recommend the use of differential fuel taxes.

3. (a) This statement is incorrect. The correct statement is . . . to ensure that a journey would not be made if the cost of the journey imposed on others was greater than the benefits obtained by the tripmaker.

(b) This statement is correct.

(c) This statement is incorrect. Road pricing alone will not produce a cure for congestion because the need for investment in urban highways will still remain.

4. When the flow on the highway is (i) in the region $0-q_1$, the additional cost of introducing an extra vehicle will be the private costs of the additional vehicle, as stated in (a) because increasing flow does not result in a decrease in speed.

When the flow on the highway is (ii) in the region q_1-q_2, the additional cost of introducing an extra vehicle will be the private costs of the additional vehicle together with an addition to the private costs of other vehicles, as stated in (b), because increasing flow results in a decrease in speed.

5. The appropriate curves are as indicated in figure 27.9.

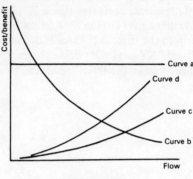

Figure 27.9

6. From equation (27.2), $v = a - bq$. From the given relationship, $v = 50 - q/100$. Hence

$$a = 50; \qquad b = \frac{1}{100}$$

From equation (27.3), $p = c + d/v$. From the given relationship, $p = 1 \cdot 25(5 + 300/v)$. Hence

$$c = 6 \cdot 25; \qquad d = 375$$

From equation 27.4

$$\text{marginal cost} = c + \frac{d}{v} \times \frac{a}{v}$$

$$= 6 \cdot 25 + \frac{18750}{v^2}$$

Flows, speeds, private and marginal costs are tabulated in table 27.3.

TABLE 27.3

Flow (p.c.u./h)	Speed $50 - q/100$ (km/h)	Private cost $1 \cdot 25(5 + 300/v)$ (p/p.c.u. km)	Marginal cost $6 \cdot 25 + 18750/v^2$ (p/p.c.u. km)	Demand $4 \times 10^4 \times p^{-1}$ (p.c.u./h)
200	48	14·06	14·39	2844
500	45	14·59	15·51	2742
1000	40	15·63	17·34	2560
1500	35	16·96	21·56	2360
2000	30	18·75	27·08	2134
2500	25	21·25	36·25	1882
3000	20	25·00	53·12	1600
4000	10	43·75	193·75	914

The private cost, marginal cost and demand curves are plotted and the intersection of the private cost and demand curves gives (a) the flow on the highway when each tripmaker is only aware of his private costs (see figure 27.10). The intersection of the marginal cost and demand curves gives (b) the flow on the highway when a trip is only made if the benefit of a trip to the tripmaker exceeds the additional cost imposed on other tripmakers. The ordinate XY gives the pricing charge that will result in flow condition (b). With the flows known the travel speeds can be calculated from equation 27.2, and are given below.

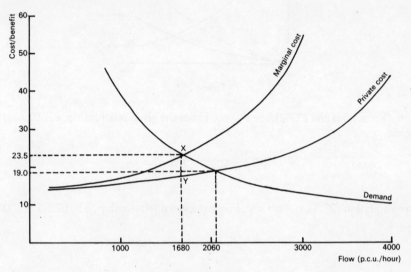

Figure 27.10

From the graphical plot:

flow condition (a)

$$q = 2060 \text{ p.c.u./h}$$

From equation 27.2

$$v = 29 \cdot 40 \text{ km/h}$$

Flow condition (b)

$$q = 1700 \text{ p.c.u./h}$$

from equation 27.2

$$v = 33 \cdot 00 \text{ km/h}$$

(c) the pricing charge

$$XY = 5 \cdot 9 \text{ p/km}$$

PART 3[†]

TRAFFIC SIGNAL CONTROL

[†] See the Appendix, page 370, for a definition of the Symbols used in part 3.

28

Introduction to traffic signals

It has been stated that the first traffic signal to be installed in Great Britain was erected in Westminster in 1868. It was illuminated by town gas and unfortunately for the future development of signals of this type was demolished by an explosion. Not until 1918 were signals used again for the control of highway traffic when manually operated three-colour light signals were introduced in New York. Some seven years later manually controlled signals were used in Piccadilly, followed in 1926 by the first automatic traffic signals in Great Britain, which were erected at Wolverhampton.

In the first signals alternate red and green fixed time periods were automatically timed and while they replaced police manpower they were not as efficient as manual control because they could not respond to changes in the traffic flow. As a refinement controllers were introduced, which were able to vary the relative durations of the fixed time green and red periods according to a preset timing pattern. It was thus possible to operate the signals with different sequences for the morning, mid-day and evening peak periods.

Where a major highway had several intersections along its length each controlled by traffic signals then means of allowing a nearly continuous progression of traffic along the major route were developed by linking the signals using a master timing device or controller instead of individual timing devices at each intersection.

During the early 1930s there was a further attempt to increase the ability of the signal controllers to deal with varying traffic demand by the incorporation of systems that would allow the signals to respond to individual vehicles. In some situations drivers were requested to sound their horns into microphones at the sides of the highway and later electrical contacts were operated by the passage of vehicles.

The vehicle detection method that became widely adopted in this period was the pneumatic tube detector and this survived in various forms in Great Britain until the 1960s. It had the disadvantages of being easily damaged by vehicles, particularly during snow-ploughing operations, and as these detectors were set in a concrete base obstructions to traffic were caused during installation and maintenance. They have now been replaced by the inductance detector, a cable set into the road surface that detects the passage or presence of a vehicle by a change in the electric field. The standard method of detection in the United Kingdom is known as 'System D' where detection for a green signal indication is carried out by a buried loop positioned 40 metres from the stop line. Other loops to extend the existing green indication are sited between the initial loop and the stop line at distances of 12 metres and

269

25 metres from the stop line.

More recently, micro-wave detectors commonly seen on temporary traffic signals at roadworks have been introduced for permanent traffic signal installations. They offer the opportunity for lower installation costs and less traffic delay during any subsequent maintenance.

Problem

Which of the following signal timing devices or controllers would operate most efficiently when the traffic flows on the intersection-approach highways vary considerably:

(a) an isolated intersection where the signals are operated by a fixed-time controller;

(b) an intersection where the signals are operated by a fixed-time controller, which also controls adjacent intersections so as to assist progression along the major route;

(c) an intersection where the signals operate by vehicle actuation using inductance loops beneath the road surface?

Solution

(a) Incorrect. At isolated intersections where the signals are operated by a fixed-time controller the durations of the red and green periods are fixed with respect to average traffic flows. They are for this reason insensitive to short-term fluctuations in the traffic flow pattern.

(b) Incorrect. At an intersection where the signals are operated by a fixed-time controller, which also controls adjacent intersections so as to assist progression along the major route, major road vehicles are not unduly delayed when the traffic flows are similar to those for which the durations of the green and red periods were calculated. When the traffic flows fluctuate on a short-term basis these fixed red and green periods cause excessive delays to both major and minor road vehicles.

(c) Correct. At an intersection where the signals are actuated by vehicles passing over pneumatic or inductance detectors the signals respond to the vehicles on the traffic signal approaches and so are able to control traffic efficiently even when there are short-term variations in the flow.

29

Warrants for the use of traffic signals

Current criteria for the installation of traffic signals in Great Britain are contained in Technical Memorandum H1/73 issued by the Department of the Environment. Each intersection is considered on the basis of traffic flow, pedestrian safety, accident record and traffic conflicts, and savings in manpower due to the elimination of manual control.

Where the economic benefits of the installation of traffic signals are being evaluated then the saving by the elimination of manual control and from expected reductions in accidents can be easily valued. As it is usual for traffic signals to be erected at junctions where priority control has previously been employed, delays at, and capacities of signal and priority control are frequently compared.

For priority control of junctions, that is the use of 'STOP' and 'GIVE WAY' commands on the minor road, delays and capacities are usually estimated using an analytical solution to the problem derived by J. C. Tanner[1].

In this analysis the traffic movements at the intersection were simplified to a single major-road stream intersected by a single minor road stream, the intersection movement taking the form of a merge or a crossing movement.

The factors affecting delay to minor road drivers in this analysis are:

1. q_1 the major road flow.
2. q_2 the minor road flow.
3. B_1 the mean time headway between bunched vehicles on the major road.
4. B_2 the mean time headway between vehicles as they discharge from the minor road without conflict with the major road flow.
5. α the mean lag or gap in the major road stream that is accepted by minor road drivers.

Values of B_1, B_2 and α will vary with site conditions particularly with the visibility of the major road from the minor road. Figure 29.1, reproduced from Road Research Technical Paper No. 56, 'Traffic Signals'[2] shows how average delay to all vehicles for a T-junction under priority control varies with total flow entering the junction. Two curves are shown for Type A conditions where visibility is good and Type B conditions where visibility is poor. It is assumed that at this junction the ratio of major road to

271

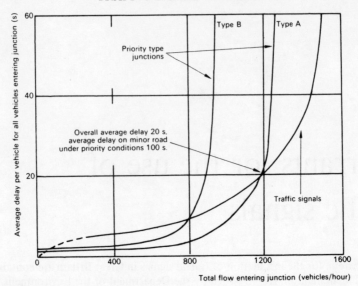

Figure 29.1 Theoretical delay/flow curves for a hypothetical T-junction under priority and signal control (based on ref. 2)

minor road flow is 4 : 1. This means that an average delay of 45 s to a minor road vehicle gives an average delay to all vehicles entering the intersection of 9 s.

Also shown on figure 29.1 is the curve of average delay to all vehicles entering the intersection when traffic signal control is used. The calculation of delay at traffic signals is discussed subsequently in chapter 43.

From a consideration of figure 29.1 it can be seen that for the values of B_1, B_2 and α selected for the priority intersection and, for the flow values selected for the traffic-signal approaches, traffic signals would be justified when the total flow entering the junction exceeds 800 veh/h or 1200 veh/h, according to visibility conditions.

The installation of traffic signals will result in a reduction of the probable points of conflict between traffic streams compared with priority control. The conflict points are illustrated in figure 29.2.

The conflicts between these two types of control are tabulated in table 29.1.

The effect of these reductions in conflicts can be illustrated by the observed accident changes at 21 sites when the form of control was varied. It was reported by F. Garwood and J. C. Tanner in a paper, 'Accident studies before and after road changes'[3], that there was a 40 per cent reduction in accidents when signal control replaced priority control.

TABLE 29.1

Conflict type	Priority	Traffic signal
crossing	16	2
merging	8	2
diverging	8	4

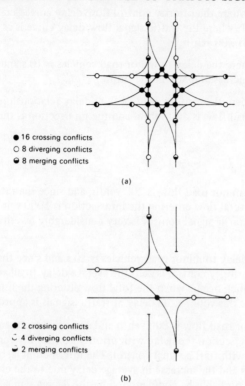

16 crossing conflicts
8 diverging conflicts
8 merging conflicts

(a)

2 crossing conflicts
4 diverging conflicts
2 merging conflicts

(b)

Figure 29.2 Traffic conflicts at priority and signal control (a) conflicts at priority control (b) conflicts at signal control

Because of the many factors involved in the calculation of delays and capacities at intersections with traffic signal and priority control each intersection must be considered individually.

References

1. J. C. Tanner. A theoretical analysis of delays at an uncontrolled intersection. *Biometrika*, **49** (1962), 163–70
2. Road Research Laboratory. Traffic Signals. *Tech. Pap. Rd Res. Bd* 56 H.M.S.O. (1966)
3. F. Garwood and J. C. Tanner. Accident studies before and after road changes. Public Works Municipal Services Congress (1956). Final report pp. 329–54

Problem

Which of the following traffic situations at an intersection would be likely to warrant the change of control from priority to signal. Indicate your answer as, very likely; likely; not likely; or extremely unlikely.

(a) An intersection where the priority control flow/delay curve is represented by curve B of figure 29.1, where the traffic signal flow/delay curve is as shown, and where the minor road flow is 200 veh/h.

(b) As above, but where the delay to minor road vehicles is 10 s and traffic volumes are not expected to increase rapidly in the future.

(c) As above, but where the priority control flow/delay relationship is illustrated by curve A, the major road flow is 800 veh/h, and the intersection is the scene of frequent traffic accidents.

Solution

(a) Very likely. The minor road flow is 200 veh/h, and since the ratio of major/minor road flow is 4 : 1 the total flow entering the intersection is 1000 veh/h. From figure 29.1 the delay with traffic signal control is very considerably less than with priority control.

(b) Not likely. The delay to minor road vehicles is 10 s and since the delay to major road vehicles under priority control is zero the average delay to all vehicles entering the intersection is 2 s, which occurs when the total flow entering the intersection is approximately 400 veh/h. At this volume the delay at traffic signals is approximately 5 s.

(c) Likely. The major road flow is 800 veh/h and since this represents 0·8 of the total flow entering the intersection the delay with priority control for conditions of good visibility is less than with traffic signal control. The intersection is, however, the scene of frequent accidents and the increase in average delay due to the change to traffic signal control would probably be compensated by the decrease in accidents. As the traffic volumes are increasing rapidly, the change to signal control would be justified on delay considerations after a short period of time.

30

Phasing

In the control of traffic at intersections the conflicts between streams of vehicles are prevented by a separation in time. The procedure by which the streams are separated is known as phasing. A phase has been defined as the sequence of conditions applied to one or more streams of traffic, which during the cycle receive simultaneous identical signal indications.

The selection and use of phases is conveniently illustrated by the conventional cross roads where the major conflicts are between the north–south traffic stream and the east–west traffic stream. Because there are two major traffic conflicts then they may be resolved by two phases. The traffic movements during each of the phases are illustrated by figure 30.1.

In some intersection traffic situations there are more than two major traffic conflicts and then it is necessary to employ more than two phases in the traffic-control system. A typical situation is where at a normal cross roads there is a heavy right-turning movement on one of the approaches. There are now three major traffic conflicts and they may be resolved by the use of a three-phase control system. The traffic movements in this instance are illustrated by figure 30.2.

The number of phases employed at any intersection are kept to a minimum compatible with safety. The reason for this is explained subsequently in chapter 31.

There are two alternative ways in which the control of traffic movements at signals can be described. Phase control which refers to the periods of green and red time allocated to each traffic stream, and Stage control which refers to the sequential steps in which the junction control is varied. In figure 30.1 phase A refers to the signal indications that control the N/S traffic stream and associated turning movements whilst phase B refers in a similar manner to the E/W traffic stream. In figure 30.2 phases A, B and C refer to the signal indications, which control the N/S, W and E traffic streams respectively. In figure 30.3 the sequence of green/red signal indications control traffic on the E/W road is referred to as phase A whilst similarly phase B refers to the signal indications for the S approach.

On the other hand a stage usually commences from the start of an amber period and always ends at the start of the following stage. A stage usually, but not always, contains a green period and stages are arranged to follow each other in a predetermined order, although they may be omitted to reduce delay. In figure 30.1 stage 1 would be the green indication for the N/S traffic movements and it would be followed by stage 2 which would be the green indication for E/W traffic movements.

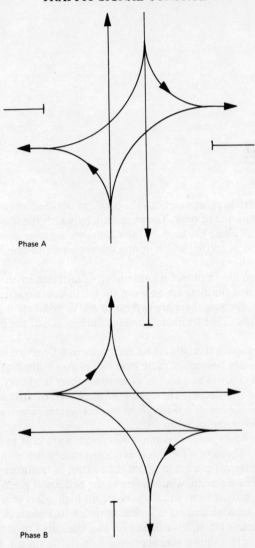

Figure 30.1 Traffic movements in a two-phase system

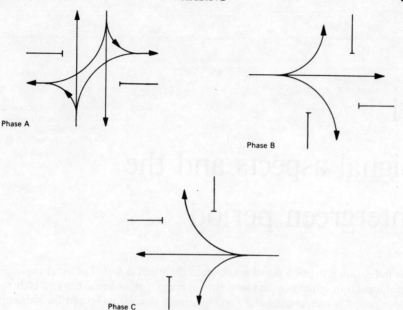

Phase A

Phase B

Phase C

Figure 30.2 Traffic movements in a three-phase system

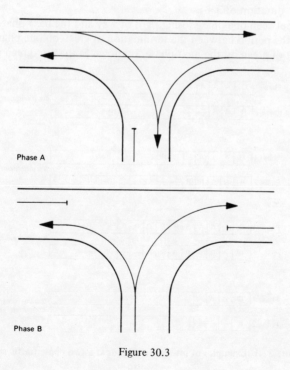

Phase A

Phase B

Figure 30.3

31

Signal aspects and the intergreen period

The indication given by a signal is known as the signal aspect. The usual sequence of signal aspects or indications in Great Britain is red, red/amber, green and amber. The amber period is standardised at 3 s and in all new signal installations the red/amber at 2 s.

In some older installations the amber indication on the first phase is shown concurrently with the red/amber indication on the second phase: in this case the red/amber indication has a duration of 3 s.

The period between one phase losing right of way and the next phase gaining right of way, that is the period between the termination of green on one phase and the commencement of green on the next phase, is known as the intergreen period.

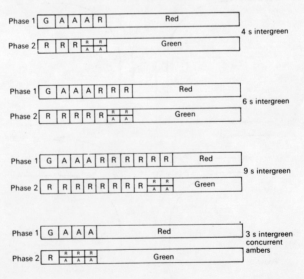

Figure 31.1 Examples of intergreen periods at a two-phase traffic signal

278

With modern controllers the minimum intergreen period is 4 s. Where right-turning vehicles are likely to become trapped in the middle of the intersection during the intergreen period, then the intergreen period may be increased in duration. It is also increased when there is a likelihood of vehicles crossing the stop line at the commencement of the red period due to high approach speeds. In this case the intergreen period results in a greater margin of safety.

Varying intergreen periods are illustrated in figure 31.1.

Intergreen periods are kept to a minimum consistent with safety because in every intergreen period there is a loss of running time for both approaches where the signal aspects are changing.

Referring to figure 31.1 it can be seen that when the intergreen period is 4 s there is a period of 1 s within the intergreen period when the signal indication is not amber on either approach. As vehicles may cross the stop line under certain circumstances during the amber period then this is a period of 1 s during which traffic movement is not possible on either approach. For this reason this period is known as lost time during the intergreen period. It should not be confused with lost time due to starting delays, which is dealt with subsequently in chapter 35.

Problems

(a) A driver approaching a traffic signal with a green indication can expect the minimum time that will elapse before the signal changes to red to be

4 s,   3 s,   2 s

(b) A driver waiting at the stop line of a traffic signal with a red indication can expect the minimum time that will elapse before the signal changes to green to be

4 s,   3 s,   2 s

(c) When the intergreen period is 7 s the period between the last possible vehicle crossing the stop line on the approach losing right of way and the first possible vehicle crossing the stop line on the approach gaining right of way is

4 s,   3 s,   2 s

(d) When the intergreen period is 7 s the lost time in the intergreen period is

4 s,   3 s,   2 s

Solutions

(a) The minimum period between a green and a red indication is the 3 s period of amber.

(b) The minimum period between a red and a green indication is the 2 s period of red/amber.

(c) A vehicle may cross the stop line at amber if it cannot stop with safety and hence the period between a vehicle crossing the stop line on the approach losing right of way and a vehicle crossing the stop line on an approach gaining the right of way is $7 - 3 = 4$ s.

(d) When the intergreen period is 7 s, the lost time in the intergreen period is 4 s.

32

Vehicle-actuated signal facilities

Fixed time operation of signals where the cycle time and green time splits do not alter during the day is usually an unsatisfactory method of control. It produces long delays to vehicles travelling through the intersection and frustration to drivers who are held at a red signal without any conflicting traffic streams being apparent.

For these reasons the control of the signals is usually varied by one of the following methods:

(a) By instruction given by an integral group timer known as cableless linking.
(b) By instruction from a central computer.
(c) By vehicle actuation where the signal indicators respond to the detection of vehicles.

Co-ordination of two or more junction controllers is possible by employing units that are synchronised by the mains supply frequency and that also incorporate a solid-state memory store. In this memory can be stored details of differing control plans for the signalled junctions. According to the time of day or day of the week the control plan appropriate for the expected traffic conditions can be selected. Future variations in traffic flow should of course be predictable if the control plans are to perform satisfactorily.

Traffic signals over a wide area can also be controlled by instructions from a central computer. Cycle time, the start and the length of the green stage together with the offset between green stages at adjacent controllers are controlled by the computer from traffic plans stored in the computer and based on past patterns of traffic flow. Traffic plans are changed according to the time of day and/or the day of the week. Area-wide control of traffic by computer is considered subsequently in greater detail.

If vehicle actuation is employed then a series of buried loops are placed on the approaches with the initial detector some 40 metres distant from the stop line. When this method of control is used then the method of control is varied in the following manner.

When a green signal is displayed it is desirable for the green indication to be shown for an initial fixed period which cannot be overriden by other demands. Such a period is built into traffic signal controllers, its value normally being 7 seconds.

Exceptions are when pedestrians walk with green on a stage that does not allow vehicles to cross their path, when the minimum green must be related to the pedestrian crossing time. When the proportion of heavy vehicles on an approach is high and they accelerate slowly owing to the gradient, then the minimum green time may be increased. On late start and early cut-off stages shorter minimum green times are frequently used.

A vehicle detected on an approach during the display of the green indication will normally extend the period of green so that a vehicle can cross the stop line before the expiry of green. Usually there are three loops on an approach, each one of which extends the green time by 1.5 seconds. If the approach has a steep gradient a longer extension period will be required.

When a vehicle is detected approaching a red signal indication the demand for the green signal is stored in the controller which serves stages in cyclic order and omits any stages for which a demand has not been received.

The demand for the green stage is satisfied when the previous stage that showed a green indication has exceeded its minimum green period and there has not been a demand for a green extension on the running stage or the last vehicle extension on the running stage has elapsed and there has not been a further demand. Alternatively, the demand for the green stage is satisfied if, after the demand is entered in the controller, the running stage runs for a further period of time known as the maximum green time. This would occur if there were continuous demands for green on the running stage. The first type of change is known as a gap change and is safer than the second type which is known as a maximum green change.

Calculation of the maximum green time depends on traffic flow conditions and is described subsequently.

Problem

For the effects a, b, c, d given select the appropriate causes.

Effects:

(a) The red signal indication on a traffic-signal approach changes to red/amber.
(b) The green signal indication on a traffic-signal approach changes to amber.
(c) The green signal indication on a traffic-signal approach remains at green.
(d) The red signal indication on a traffic-signal approach remains at red.

Causes:

(e) The green indication on the preceding phase has run to maximum green and there has been a demand for green on the approach being considered.
(f) There has been a demand for the extension of the green period on the preceding phase and that phase has not yet reached its maximum green period.
(g) The minimum green period on the approach being considered has not as yet expired.
(h) The last vehicle extension on the phase being considered has expired and there has been a demand for the green signal on the next phase.

Solution

The effect (a) is the result of cause (e).
The effect (b) is the result of cause (h).
The effect (c) is the result of cause (g).
The effect (d) is the result of cause (f).

33

The effect of roadway and environmental factors on the capacity of a traffic-signal approach

The capacity of a traffic-signal controlled intersection is limited by the capacities of the individual approaches to the intersection.

There are two types of factor which affect the capacity of an approach: roadway and environmental factors, discussed in this chapter, and traffic and control factors discussed in chapter 34.

The roadway and environmental factors that control the capacity of an approach are the physical layout of the approach, in particular its width, the radii along which left- or right-turning vehicles have to travel, and the gradient of the approach and its exit from the intersection.

The capacity of an approach is measured independently of traffic and control factor and is expressed as the saturation flow.

Saturation flow is defined as the maximum flow, expressed as equivalent passenger cars, that can cross the stop line of the approach when there is a continuous green signal indication and a continuous queue of vehicles on the approach.

Observations of traffic flow made by the Road Research Laboratory at intersections in the London area and also in some of the larger cities, supplemented by controlled experiments at the Laboratory test track, have shown that the saturation flow(s) expressed in passenger car units per hour with no parked vehicles is given by

$$s = 525w \text{ p.c.u./h}$$

where w is the width of the approach in metres.

This formula is applicable to approach widths greater than 5·5 m; at widths less than 5·5 m the relationship is not linear and saturation flows may be estimated from table 3

TABLE 33.1

w (m)	3	3·5	4	4·5	5	5·5
s (p.c.u./h)	1850	1875	1975	2175	2550	2900

These saturation flows have to be amended for the effect of gradient. This modification has been reported as a decrease or increase of 3 per cent in the saturation flow for every 1 per cent of uphill or downhill gradient of the approach. The gradient of the approach was defined as the average slope between the stop line and a point on the approach 61 m before it. At the sites where these observations were made, the slope continued through the intersection.

Where vehicles crossing the stop line have then to travel immediately around a curve the rate of discharge across the stop line will be reduced. This occurs frequently when right-turning vehicles are able to discharge during a right-turning phase. Test-track experiments have shown that the saturation flow for right-turning streams may be obtained from

$$s = \frac{1800}{1 + 1·52/r} \text{ p.c.u./h, for single-file streams}$$

or 1600 p.c.u./h

$$s = \frac{3000}{1 + 1·52/r} \text{ p.c.u./h for double-file streams}$$

or 2700 p.c.u./h

where r = turning radius in metres.

The environment also has an effect on the saturation flow of an approach and while it difficult to define this effect precisely, generalised modification factors are often applied.

Where a site is designed with a good environment, that is dual carriageway approaches, no noticeable pedestrian interference, no parked vehicles, no interferences to traffic flow from right-turning vehicles, good visibility and adequate turning radii then the saturation flow is taken as 120 per cent of the standard value.

If however a site is designed with poor environment, that is low average speeds, interference from standing vehicles and right-turning vehicles, poor visibility and poor alignment, then the saturation flow is taken as 85 per cent of the standard value.

Determination of the saturation flow of a traffic-signal approach

To determine the saturation flow of an approach select one in which there is a continuous queue even at the end of the green period. Avoid situations in which right-turning vehicles have an erratic effect on the traffic flow. For ease of observation it is preferable to select an approach that is restricted to straight ahead and left-turning vehicles.

Using a stopwatch note the number, type and turning movement of each vehicle crossing the stop line during each successive 0·1 minute interval of the green and amber period. At the end of the amber period there will normally be an interval of less than 0·1 minute. Note the length of this interval and also the number and type of vehicles crossing the stop line in the interval. These intervals are subsequently referred to as last saturated intervals.

If at any time the flow on the approach is not saturated, then observations should be discontinued until the flow reaches saturation level again.

If it is not found possible to observe vehicle type then only the number and turning movement of vehicles should be noted. At the completion of observations a separate count is then necessary to determine the composition of the traffic flow.

The observations given in table 33.2 were obtained at a traffic-signal controlled intersection in the City of Bradford. In this instance observations were made of mixed vehicles travelling straight ahead.

TABLE 33.2　Observed discharge of vehicles across the stop line

Time (minute)	0	0·1	0·2	0·3	0·4	0·5
No. of vehicles crossing stopline	60	76	71	78	79	
No. of saturated intervals observed	32	32	32	32	32	
Discharge per 0·1 min		1·88	2.38	2·22	2·44	2·47

Total duration of the last saturated intervals　　　= 142 seconds
Total number of vehicles crossing stop line　　　= 41.
Discharge per 0·1 minute during last saturated interval = 41 × 6/142
　　　　　　　　　　　　　　　　　　　　　　　　= 1·74 vehicles/s

During the first and last saturated intervals there is a loss of capacity because of the effect of vehicles accelerating from the stationary position at the commencement of the green period and decelerating during the amber period.

The flow during the remainder of the observed periods represents the maximum discharge possible and their mean value gives the saturation flow for the approach

$$\text{saturation flow} = \frac{2\cdot48 + 2\cdot22 + 2\cdot44 + 2\cdot47}{4}$$

$$= 2\cdot40 \text{ vehicles per } 0\cdot1 \text{ minute}$$

$$= 1440 \text{ vehicles/h}$$

This value must now be converted to traffic-signal passenger car units and a subsidiary traffic count is required to determine the composition of the traffic.

Observe the composition and turning movements of the traffic flow for a period of 30 minutes and at similar time to when the original observations were made. The following composition of traffic was noted on the approach where the flow figures given in table 33.2 were observed (using the equivalent effects of various vehicle types given in chapter 34)

　　　　　　　　heavy vehicles　　　　14 per cent
　　　　　　　　buses　　　　　　　　5 per cent
　　　　　　　　motor cycles　　　　　6 per cent
　　　　　　　　private cars　　　　　75 per cent
　　　　　　　　all vehicles proceed straight ahead

The passenger car equivalent of the flow is then

$$0.14 \times 1.75 + 0.05 \times 2.25 + 0.06 \times 0.33 + 0.75 \times 1 = 1.16$$

Saturation flow = 1440 x 1.16
           = 1670 p.c.u./h.

The design figure given in *Road Research Technical Paper* 56 is 1900 p.c.u./h for a 3.65 m lane width.

Problems

Four differing traffic signal approaches are described below. Place them in the order of their traffic capacity.

(a) An approach with good environmental conditions where all vehicles discharge straight across the intersection and where the approach width is 7.30 m.

(b) An approach with poor environmental conditions with a continuous uphill gradient of 3 per cent, where all vehicles discharge straight across the intersection and where the approach width is 10.50 m.

(c) An approach with normal environmental conditions from which all vehicles turn right in a double-file stream on a path with a radius of 30 m.

(d) An approach with good environmental conditions and downhill gradient of 4 per cent, where all vehicles discharge straight across the intersection and where the approach width is 5.20 m.

Solutions

The traffic capacities of the approaches are:

(a) saturation flow = 525 x 7.3 = 3833 p.c.u./h
plus environmental factor of 25 per cent

$$= 3833 \times 1.2$$
$$= 4600 \text{ p.c.u./h}$$

(b) saturation flow = 525 x 10.5 = 5513 p.c.u./h
minus environmental factor of 15 per cent

$$= 5513 \times 0.85$$
$$= 4686 \text{ p.c.u./h}$$

minus gradient effect of 3 x 3 per cent

$$= 4686 \times 0.91$$
$$= 4264 \text{ p.c.u./h}$$

(c) saturation flow $= \dfrac{3000}{1 + 1 \cdot 52/r}$ p.c.u./h

$ = \dfrac{3000}{1 + 1 \cdot 52/30}$

$ = 2857$ p.c.u./h

(d) saturation flow $= 2700$ p.c.u./h (table 33.1)
plus environmental factor of 20%

$$= 2700 \times 1 \cdot 2$$

$$= 3240 \text{ p.c.u./h}$$

plus gradient effect of 4 x 3%

$$= 3240 \times 1 \cdot 12$$

$$= 3629 \text{ p.c.u./h}$$

The order of capacity of the approaches is (a), (b), (d), (c).

34

The effect of traffic factors on the capacity of a traffic-signal approach

The effect of traffic factors on the capacity of an approach is allowed for by the use of passenger car units, which represent the effect of varying vehicle types relative to the passenger car.

These equivalents are given in table 34.1.

TABLE 34.1

1 bus	2·25 p.c.u.
1 heavy or medium goods vehicle	1·75 p.c.u.
1 light goods vehicle	1·00 p.c.u.
1 motor cycle, moped or scooter	0·33 p.c.u.
1 pedal cycle	0·20 p.c.u.

Where right-turning vehicles are mixed with straight-ahead vehicles on an approach, the saturation flow on that approach is calculated as if all vehicles travelled straight ahead but the passenger car equivalents of right-turning vehicles are increased by 75 per cent.

The number of vehicles crossing the stop line in a given period of time is dependent not only on the saturation flow, but also on the proportion of the time during which the signal is effectively green (λ).

A cycle is a complete sequence of signal indications, that is a green period and a red period for a two-phase system and the time during which the signal is effectively green during a cycle, known as the effective green time. The maximum number of vehicles crossing the stop line is then

$$\frac{\text{saturation flow} \times \text{effective green time}}{\text{cycle time}} \text{ p.c.u./h} = \text{saturation flow} \times \lambda$$

Determining the passenger car equivalent of a goods vehicle in single-lane flow at traffic signals

It has been noted that the passenger car equivalent of a goods vehicle varies with the approach width and it has been suggested that a lower value of the p.c.u. equivalent should be taken for single-lane flow.

Controlled experiments carried out by the Road Research Laboratory[1] showed that for a 3·05 m approach width the passenger car equivalent of a goods vehicle was 1·47, and for a 3·65 m approach width a value of 1·68 was obtained.

The passenger car equivalent of a goods vehicle may be determined by noting the time headway between successive vehicles as they cross the stop line.

Select a saturated approach where traffic is confined to a single lane and all vehicles continue straight across the intersection.

Note the time headway between individual pairs of vehicles, dividing observations into the following groups:

1. A car following a car.
2. A car following a goods vehicle.
3. A goods vehicle following a car.
4. A goods vehicle following a goods vehicle.

Vehicles crossing the stop line within three seconds of the commencement and termination of the green period should not be included in the observations because of the effects of acceleration and deceleration.

Observations obtained at a traffic-signal approach in the City of Leeds are shown in table 34.2.

TABLE 34.2

Approach width (m)		Car following car	Car following goods	Goods following car	Goods following goods
3·65	no. of headways	66 (*a*)	33 (*b*)	32 (*c*)	9 (*d*)
	mean headway (seconds)	1·9 (*w*)	2·8 (*x*)	2·3 (*y*)	3·0 (*z*)

The passenger car unit for a goods vehicle can be found by dividing the mean headway for a goods vehicle following a goods vehicle by the mean headway for a car following a car. This is true if the effect of a goods vehicle is independent of whether the vehicles preceding it and following it are light or heavy. The necessary and sufficient condition for this is that the sum of the average headways for car following car and goods vehicle following goods vehicle should be equal to the sum of the average headways for a car following a goods vehicle and a goods vehicle following a car.

If this condition is not met then corrected values of the mean headways can be calculated as below.

Corrected value of mean headway for a car following a car
= Uncorrected value (w) − correction/no. of headways of this type (a)
Corrected value of mean headway for a car following a goods vehicle
= Uncorrected value (x) + correction/no. of headways of this type (b)
Corrected value of mean headway for a goods vehicle following a car
= Uncorrected value (y) + correction/no. of headways of this type (c)
Corrected value of mean headway for a goods vehicle following a goods vehicle
= Uncorrected value (z) − correction/no. of headways of this type (d)

where

$$\text{correction} = \frac{abcd(w - x - y + z)}{bcd + acd + abd + abc}$$

Using the values given in table 34.2 the correction is calculated. Corrected values of the mean headways are given in table 34.3.

TABLE 34.3

	Car following car	Car following goods	Goods following car	Goods following goods
Adjusted mean headway (seconds)	1·9	2·7	2·3	3·1

The passenger car equivalent is then 1·6, which can be compared with the usual value of 1·75 and values of 1·47 and 1·68 for approach widths of 3·05 m and 3·65 m, respectively, determined in reference 1.

Reference

1. Road Research Laboratory. Determination of the passenger car equivalent of a goods vehicle in single-lane flow at traffic signals. *Rd Res. Laboratory Report* LN/573/DAS (1964)

Problem

The hourly traffic flow on a signal approach is composed of 400 passenger cars, 200 heavy goods vehicles and 60 motor cycles. One-fifth of all the vehicles turn right. The width of the approach is 5·5 m. If the approach is just able to pass all the traffic, should the ratio of effective green time to cycle time be 0·29, 0·33 or 0·55?

Solution

Converting the flow to passenger car units

$$320 \text{ straight-ahead cars} \quad = 320 \times 1 \quad = 320 \text{ p.c.u.}$$

$$80 \text{ right-turn cars} \quad = 80 \times 1 \cdot 75 \quad = 140 \text{ p.c.u.}$$

$$160 \text{ straight-ahead goods} = 160 \times 1 \cdot 75 = 280 \text{ p.c.u.}$$

$$40 \text{ right-turn goods} \quad = 40 \times 1 \cdot 75$$

$$\times 1 \cdot 75 = 122 \text{ p.c.u.}$$

$$48 \text{ straight-ahead m/c} \quad = 40 \times 0 \cdot 33 \quad = 13 \text{ p.c.u.}$$

$$12 \text{ right-turn m/c} \quad = 12 \times 0 \cdot 33$$

$$\times 1 \cdot 75 = 7 \text{ p.c.u.}$$

$$\overline{}$$

$$882 \text{ p.c.u.}$$

$$\text{maximum flow across stop line} = \frac{\text{saturation flow} \times \text{effective green time}}{\text{cycle time}}$$

$$882 = \frac{3025 \times \text{effective green time}}{\text{cycle time}}$$

$$\frac{\text{effective green time}}{\text{cycle time}} = \frac{882}{3025} = 0 \cdot 29$$

35

Determination of the effective green time

In chapter 34 the concept of effective green time was introduced as a means of determining the number of vehicles that could cross a stop line over the whole of the cycle comprising both red and green periods.

It is obvious that in practice the flow across the stop line cannot commence or terminate instantly for there is a flow and at the end of the green indication it is slowly reduced to zero.

A study of the discharge of vehicles across the stop line allows the effective green time to be determined. Figure 35.1 shows the variation of discharge with time, the area beneath the curve representing the number of vehicles that cross the stop line during the green period. The area beneath the curve is not easily determined and for convenience a rectangle of equal area to that under the curve is superimposed upon the curve. The height of the rectangle is equal to the saturation flow and the base of the rectangle is the effective green time.

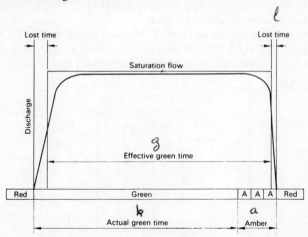

Figure 35.1 Variations in the discharge across the stop line

291

As the amber indication is a period during which, under certain circumstances, vehicles may cross the stop line the discharge across the stop line commences at the beginning of the green period and terminates at the end of the amber period. The time intervals between the commencement of green and the commencement of effective gre and also between the termination of effective green and the termination of the amber period are referred to as the lost time due to starting delays.

From figure 35.1 it can be seen that the actual green time plus the amber period is equal to the effective green time plus the lost time due to starting delays.

In actual practice the lost time due to starting delays is taken as 2 s and so the effective green time is equal to the actual green time plus the 3 s amber period minus the 2 s lost time.

Determination of the lost time on a traffic-signal approach

The observations made in chapter 33 will be used to determine the lost time due to starting delays at a traffic signal-controlled intersection in the City of Bradford. The observations are repeated below in table 35.1.

TABLE 35.1

Time (minutes)	0	0·1	0·2	0·3	0·4	0·5
No. of vehicles crossing stop line	60	76	71	78	79	
No. of saturated intervals observed	32	32	32	32	32	
Discharge per 0·1 minute	1·88	2·48	2·22	2·44	2·47	

No. of last saturated intervals 24, average duration 5·9 s

Total duration of the last saturated intervals   = 142 seconds

Total number of vehicles crossing stop line   = 41

Discharge per 0·1 minute during last saturated interval = 41 × 6/142 = 1·74 vehicles

Additionally, it has been previously calculated in chapter 33 that the saturation flow was 1670 p.c.u./h or 2·40 vehicles/0·1 minute.

The lost time at the beginning and end of the green period may be calculated by reference to figure 35.2. By definition the number of vehicles represented by the rectangle efij is equal to the number of vehicles represented by the original histogram

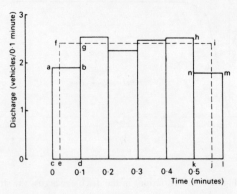

Figure 35.2 Observed discharge across the stop line

(because total flow during effective green time is equal to total flow during green plus amber period). The number of vehicles represented by the area dghk is also equal to the number of vehicles represented by the four 0·1 minute periods of saturated flow between d and k.

This means that:

1. The number of vehicles represented by abdc is equal to the number of vehicles represented by efgd.
2. The number of vehicles represented by hijk is equal to the number of vehicles represented by nmlk.

That is

$$ed \times 2 \cdot 40 = 1 \cdot 88 \times 0 \cdot 1$$

giving

$$ed = 0 \cdot 08 \text{ minute}$$

or

$$ce = 0 \cdot 02 \text{ minute}$$

similarly

$$jl = 0 \cdot 03 \text{ minute}$$

and

$$jl + ce = \text{lost time during green phase} = 2 \cdot 9 \text{ s}$$

This value is higher than the accepted value of 2 s and the saturation flow was also noted to be lower than the usual design value. These departures from the accepted values were considered to be caused by a high proportion of elderly drivers in the traffic flow.

Problem

The lost time due to starting delays on a traffic signal approach is noted to be 3 s; the actual green time is 25 s. Is the effective green time 24 s, 25 s or 26 s?

Solution

$$\text{Effective green time} = 25 \text{ s} + 3 \text{ s (amber)} - 3 \text{ s (lost time)}$$
$$= 25 \text{ s}$$

36

Optimum cycle times

for an intersection

The length of the cycle time under fixed time operation is dependent on traffic conditions. Where the intersection is heavily trafficked cycle times must be longer than when the intersection is lightly trafficked.

One definition of the degree of trafficking of an approach is given by the y value, which is the flow on the approach divided by the saturation flow.

For any given traffic-flow conditions with the signals operating under fixed-time control, the duration of the cycle must affect the average delay to vehicles passing through the intersection. Where the cycle time is very short, the proportion of the cycle time occupied by the lost time in the intergreen period and by starting delays is high, making the signal control inefficient and causing lengthy delays.

When on the other hand the cycle time is considerably longer, waiting vehicles will clear the stop line during the early part of the green period and the only vehicles crossing the stop line during the latter part of the green period will be those that subsequently arrive, often at extended headways. As the discharge rate or saturation flow across the stop line is greatest when there is a queue on the approach this also results in inefficient operation.

As a result of the computer simulation of flow at traffic signals carried out by the Road Research Laboratory[1], it was possible to show these variations of average delay with cycle time that occur at any given intersection when the flows on the approaches remain constant. They are illustrated in figure 36.1 and can be explained for most practical purposes by the total lost time and the y value found at the intersection.

It is shown in *Road Research Technical Paper* 39 that a sufficiently close approximation to the optimum cycle time C_0 could be obtained by the use of the following equation

$$C_0 = \frac{1 \cdot 5L + 5}{1 - Y}$$

where L is the total lost time per cycle

294

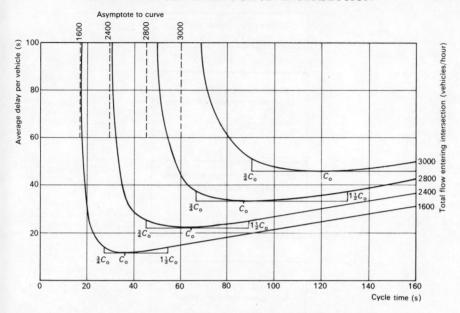

Figure 36.1 Effect on delay of variation of the cycle length (based on ref. 1)
2-phase, 4-arm intersection, equal flows on all arms, equal saturation flows of 1800 vehicles/hour, equal green times, total lost time/cycle 10 s

Y is the sum of the maximum y values for all the phases comprising the cycle.

The application of this formula may be illustrated from figure 36.1 and the calculation of the optimum cycle time C_0 is tabulated in table 36.1 for each of the four flow rates illustrated. An inspection of figure 36.1 shows that the values by the approximate formula are similar to those obtained by simulation.

TABLE 36.1

(1) Total flow (veh/h)	(2) Flow/approach (veh/h)	(3) y value (2)/1800	(4) L	(5) Y ($n \times$ (3)*)	(6) C_0 (s)
3000	750	0·42	10	0·84	125
2800	700	0·39	10	0·78	91
2400	600	0·33	10	0·66	59
1600	400	0·22	10	0·44	36

* n = number of phases.

There is a minimum cycle time of 25 s fixed from safety considerations while a maximum cycle time of 120 s is generally considered desirable. Calculated optimum cycle times should normally be limited to this range of values.

The calculation of the optimum cycle time consists of several steps, best illustrated by the flow chart given in figure 36.2.

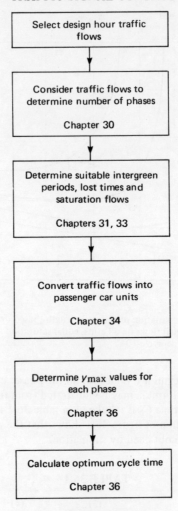

Figure 36.2

Reference

1. Road Research Laboratory. Traffic signal settings. *Tech. Pap. Rd Res. Bd* 39, H.M.S.O. (1961)

Problems

Design-hour traffic flows at a 4-arm 3-phase intersection are given in table 36.2; the intergreen period is 5 s; starting delays are 2 s per green interval. Is the optimum cycle time, 88 s, 105 s, or 115 s?

TABLE 36.2   Design-hour traffic flows

	Passenger cars	Medium and heavy goods	Buses	Motor cycles	Approach width (m)
North approach, straight ahead and left turning	500	100	10	20⎫	
North approach, right turning	50	10	0	10⎭	7·65
South approach, straight ahead and left turning	400	150	0	30⎫	
South approach, right turning	40	30	0	5⎭	7·65
West approach, straight ahead and left turning	400	50	5	10	3·65
West approach, right turning	300	60	0	20	3·65
East approach, straight ahead and left turning	200	180	4	15	3·65
East approach, right turning	260	20	0	10	3·65

Solutions

For the traffic flows given in table 36.2 the three major traffic movements are:

north/south, all directions of movement;
east/west, straight ahead and left-turning movements;
east/west, right-turning movements.

These three major traffic movements will each be given a separate phase (see page 275)

The traffic flow for each of these phases will now be converted to passenger car equivalents (table 36.3). Note that the vehicles that are given their own right-turning lane are not given a passenger car equivalent 1·75 times greater than the corresponding straight-ahead value. Instead, the reduced saturation flow applicable to a right-turning traffic lane is used.

TABLE 36.3   Design hour traffic flows in p.c.u.'s

	Passenger cars	Medium and heavy goods	Buses	Motor cycles	Saturation flow (pcu/h)
North approach, straight ahead and left turning	500	175	23	7⎫	
North approach, right turning	88	30	0	5⎭	4015
South approach, straight ahead and left turning	400	263	0	10⎫	
South approach, right turning	70	90	0	4⎭	4015
West approach, straight ahead and left turning	400	88	11	3	1900
West approach, right turning	300	105	0	7	1600
East approach, straight ahead and left turning	200	315	9	5	1900
East approach, right turning	260	35	0	3	1600

(See chapters 33 and 34.)

The y value for each approach may now be calculated (see table 36.4).

TABLE 36.4

	Approach	Flow	Saturation flow	y
1	North approach, straight ahead and left turning	828	4015	0·21
2	North approach, right turning			
3	South approach, straight ahead and left turning	837	4015	0·21
4	South approach, right turning			
5	West approach, straight ahead and left turning	502	1900	0·26
6	West approach, right turning	412	1600	0·25
7	East approach, straight ahead and left turning	529	1900	0·28
8	East approach, right turning	298	1600	0·19

(See chapters 33 and 36.)

It is now necessary to select the maximum y values for the three phases. During phase 1 traffic on lanes 1 and 2, 3 and 4 will flow, right-turning vehicles on both these approaches being mixed with straight-ahead and left-turning vehicles. The maximum value of y is on the north/south phase and is 0·21.

During phase 2 traffic on lanes 5 and 7 will flow and y_{max} occurs on the west approach; it is 0·28.

During phase 3 traffic on lanes 6 and 8 will flow and y_{max} occurs on the west approach; it is 0·25 (see page 294).

The total lost time during the cycle is composed of the lost time during the green period and also the lost time during the intergreen period.

The lost time during the 5 s intergreen period is 2 s and the intergreen period occurs three times with a three-phase system (see chapter 31).

Starting delays are 2 s for each of the three green periods of the three-phase cycle (see chapter 33).

The total lost time is $3 \times 2 + 3 \times 2 = 12$ s.

The optimum cycle time C_0 is given by:

$$C_0 = \frac{1·5L + 5}{1 - Y}$$

$$= \frac{1·5 \times 12 + 5}{1 - (0·21 + 0·28 + 0·25)}$$

$$= 88 \text{ s}$$

37

The timing diagram

Previously that optimum cycle time has been calculated which would result in minimum overall delay when employed with fixed-time signals. The intergreen periods have also been selected previously and the remaining calculation, before the whole sequence of signal aspects can be described, is to calculate the duration of the green signal aspects.

Once the duration of these green aspects has been calculated, they can be employed with vehicle-actuated signals, and are then the maximum green times at the end of which a phase change will occur regardless of any demands for vehicle extensions.

The first step is to calculate the amount of effective green time available during each cycle from

available effective green time/cycle = cycle time − lost time/cycle (L)

This available effective green time is then divided between the phases in proportion to the y_{max} value for each phase.

For the particular value of lost time due to starting delays associated with each green period, the actual green time can be calculated. This is the indication shown on the signal and depicted on the timing diagram.

These steps will now be illustrated by the following example.

C_0 82 s
L 12 s
y_{max} north/south phase (all movements) 0·21
y_{max} east/west phase (straight ahead and left turn) 0·26
y_{max} east/west phase (right turning) 0·25

available effective green time/cycle = 82 − 12 = 70 s

This is divided between the phases in the ratio

0·21:   0·26:   0·25

or

20·0 s,   25·0 s   and   25·0 s

When the lost time due to starting delays is 2 s then the actual green time is equal to the effective green time minus 1 s (see chapter 35).

The actual green times used are in practice

north/south traffic flow   19 s
east/west straight ahead and left turning   24 s
east/west right-turning flows   24 s

The actual green times have been rounded to the nearest second.
Figure 37.1 is a timing diagram showing these signal indications.

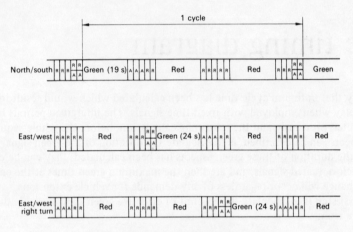

Figure 37.1 A three-phase timing diagram

Problems

An intersection is controlled by four-phase traffic signals, with a cycle time of 100 s.
The minimum intergreen period is employed and a value of lost time per green period
of 3 s is assumed. Saturation flows on all approaches are identical, but the maximum
traffic flows on two of the phases are twice the maximum traffic flows on the remaining
two phases. Which of the following series of actual green times would be appropriate:

(a) 28 s, 28 s, 14 s and 14 s;
(b) 30 s, 15 s, 30 s and 15 s;
(c) 30 s, 15 s, 15 s and 15 s;
(d) 14 s, 14 s, 30 s and 30 s.

Solutions

The minimum intergreen period that can be employed is 4 s (see page 278), and when
it has this value the lost time in the intergreen period is 1 s.
    The total lost time L per cycle is equal to the number of phases multiplied by the
lost time in the intergreen period plus the lost time caused by starting delays, that is
$4(1 + 3) = 16$ s (see page 291).

The y value for an approach is the flow divided by the saturation flow, and the maximum flows for each phase are in the ratio: $2:2:1:1$. As the saturation flows on each approach are equal, this is also the ratio of the y values.

$$\text{available effective green time/cycle} = \text{cycle time} - \text{lost time/cycle}$$
$$= 100 - 16 \text{ s}$$
$$= 84 \text{ s}$$

This available effective green time is divided in the ratio of the y values, that is

$$28\text{ s}, 28\text{ s}, 14\text{ s and } 14\text{ s}$$

$$\text{actual green time} = \text{effective green time} + \text{starting delays} - \text{amber period}$$
$$= \text{effective green time} + 3 \text{ s} - 3 \text{ s}$$

The actual green times are 28 s, 28 s, 14 s and 14 s (see page 291).

38

Early cut-off and late-start facilities

Where the number of right-turning vehicles is not sufficient to justify the provision of a right-turning phase but where right-turning vehicles have difficulty in completing the traffic movement, then an early cut-off or a late start of the opposing phase is employed.

An early cut-off of the opposing flow allows right-turning vehicles to complete their traffic movement at the end of the green period when the opposing flow is halted. To

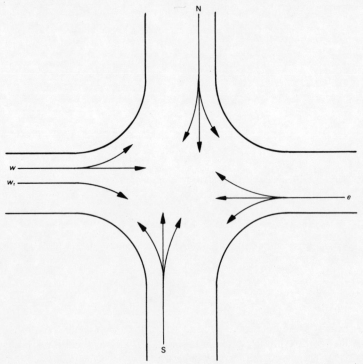

Figure 38.1 Traffic flows at an intersection with a heavy right-turning movement

allow straight-ahead vehicles on the same approach to flow without interruption during the early part of the green period, it is necessary for there to be sufficient room in the intersection for right-turning vehicles to wait.

In contrast a late-start facility discharges the right-turning vehicles at the commencement of the green period and for this reason storage space is not as important as with the early cut-off facility.

The calculation of optimum cycle times for fixed-time operation and hence the determination of maximum green times for vehicle-actuated operation is as follows.

The traffic flows are as shown in figure 38.1 and it is necessary to introduce a late-start or early cut-off facility to permit vehicles to turn right from the west approach.

These traffic flows are to be controlled by a two-phase system and it is first necessary to determine the maximum value of the ratio of flow to saturation flow for each phase.

For the north/south phase the maximum y value will be denoted by $y_{\max N/S}$. For the west/east phase, however, it is necessary to determine the greater of

$$y_w \quad \text{or} \quad y_{w_r} + y_e$$

Whichever is the greater will be the $y_{\max E/W}$.

The reason for this combination of y values is that during the west/east phase the flow w continues for the whole of the green time, while flows w_r and e share this green time.

If $y_{w_r} + y_e$ is greater than y_w, then the y value for the first stage will be y_e and that for the second stage y_{w_r}. If however y_w is the greater, then the green time that flow w requires should be divided in proportion to the y values of streams w_r and e. Then the y value for the first stage is

$$\frac{y_e y_w}{y_e + y_{w_r}}$$

and for the second stage

$$\frac{y_{w_r} y_w}{y_e + y_{w_r}}$$

The same procedure may be used with a late start for the opposing flow.

Problem

The following hourly flows (table 38.1) and saturation flows relate to an intersection to be controlled by two-phase signals incorporating a late-start feature. Minimum intergreen periods are employed and starting delays are 2 s, for each green plus amber period.

TABLE 38.1

Approach	Flow (p.c.u./h)	Saturation flow (p.c.u./h)
west, straight ahead and left turning	400	1900
west, right turning	200	1600
east, all movements	700	1900
north, all movements	500	1900
south, all movements	600	1900

Is the period that right-turning vehicles from the west approach require to complete their turning movement without obstruction from the straight ahead flow on east approach 5 s; 11 s; or 16 s?

Solution

To calculate the optimum cycle time for the intersection it is first necessary to calculate the y values for the approaches. They are tabulated in table 38.2.

TABLE 38.2

	Approach	Flow (p.c.u./h)	Saturation flow (p.c.u./h)	y value
1	west, straight ahead, left turning	400	1900	0·21
2	west, right turning	200	1600	0·13
3	east, all movements	700	1900	0·37
4	north, all movements	500	1900	0·26
5	south, all movements	600	1900	0·32

$y_{\text{max west/east}}$ is 0·21 or 0·13 + 0·37, whichever is the greater (see chapter 38).
$y_{\text{max north/south}}$ is 0·26 or 0·32, whichever is the greater.
    The total lost time is equal to the sum of the lost time in the intergreen period, which is 1 s (see page 278) plus starting delays of 2 s (see page 291) multiplied by the number of phases, that is $2(1 + 2) = 6$ s.
    The optimum cycle time C_0 is given by (see page 294),

$$C_0 = \frac{1\cdot5L + 5}{1 - Y}$$

$$= \frac{1\cdot5 \times 6 + 5}{1 - (0\cdot50 + 0\cdot32)}$$

$$= 78 \text{ s}$$

The effective green time available for distribution between the phases is the cycle time minus the total lost time per cycle (see page 292).

$$= 78 - 6$$

$$= 72 \text{ s}$$

This must be divided between the phases in the ratio of the y_{max} for each phase, that is 0·50 and 0·32 (see page 299), giving effective green times of 44 s and 28 s respectively.

The actual green time may be obtained from the effective green times by the relationship, actual green time = effective green time + starting delays − amber period (see page 291).

The actual green times are thus:

north/south phase   27 s
east/west phase   43 s

The east/west phase is divided into two stages in the proportion of 0·13 to 0·37 giving a late start to green on the east approach of 11 s. The timing diagram is shown in figure 38.2

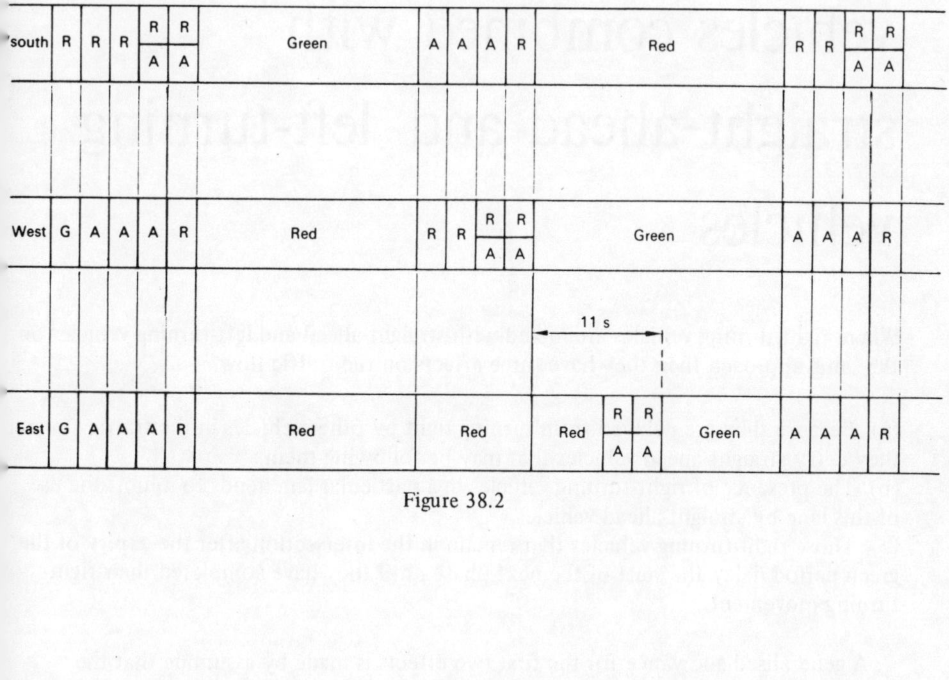

Figure 38.2

39

The effect of right-turning vehicles combined with straight-ahead and left-turning vehicles

Where right-turning vehicles are mixed with straight-ahead and left-turning vehicles on the same approach then they have three effects on the traffic flow.

(a) Because they are delayed from turning right by other vehicles in the traffic stream, they delay straight-ahead vehicles that may be following them.
(b) The presence of right-turning vehicles in a particular lane tends to inhibit the use of this lane by straight-ahead vehicles.
(c) Those right-turning vehicles that remain in the intersection after the expiry of the green period delay the start of the next phase until they have completed their right-turning movement.

A generalised allowance for the first two effects is made by assuming that the passenger car equivalent of a right-turning vehicle is 1·75 times the equivalent straight-ahead passenger car unit.

The third effect is however more difficult to estimate, for when the opposing traffic flow is not great some right-turning vehicles will be able to discharge through gaps in the opposing stream, leaving only a proportion that must discharge at the end of the green period.

At the beginning of the green period the opposing flow is discharged at the saturation rate but later in the green period when the queue has been discharged, opposing vehicles are discharged across the stop line in the random or unsaturated manner in which they arrive on the approach.

If the length of the initial saturated green time can be calculated then the length of the unsaturated green time is also known. The discharge of right-turning vehicles through gaps in the opposing flow can then be determined from the analytical work carried out by Tanner (see page 164) on the discharge of vehicles at priority intersections.

This unsaturated green period during which right-turning vehicles can be discharged through the opposing flow can be determined by assuming that no vehicles remain in the opposing queue at the end of the green period. Such an assumption is correct in the flow conditions being considered.

When the saturated green time for the opposing flow is denoted by g_s, then the number of vehicles discharged during this time is

$$(r_e + g_s)q$$

where r_e is the effective red period, that is the cycle time minus the effective green period,

q is the opposing flow.

These vehicles discharge during the saturated green time g_s at the saturation flow s. Then

$$g_s = \frac{(r_e + g_s)q}{s}$$

giving

$$g_s = \frac{r_e \times q}{s - q}$$

$$= \frac{(c - g)q}{s - q} \tag{39.1}$$

where c is the cycle time,

g is the effective green time

Let g_u be the unsaturated green time. Then

$$g_u = g - g_s$$

Substituting from equation 39.1

$$g_u = \frac{gs - qc}{s - q}$$

The number of vehicles turning right during this period n_r is then

$$n_r = s_r \left(\frac{gs - qc}{s - q} \right) \tag{39.2}$$

where s_r is the maximum theoretical right-turning flow passing through gaps in the opposing flow.

Tanner has given the maximum intersection flow s_r as

$$\frac{q_0(1 - B_0 q_0)}{\exp\left[q_0(\alpha - B_0)\right]\left[1 - \exp(-B_r q_0)\right]}$$

where q_0 is the rate of arrival of the opposing flow in vehicles per second;

B_0 is the mean minimum time interval between opposing vehicles passing through the intersection,

B_r is the mean minimum time interval between right-turning vehicles passing through the intersection.

α is the average gap accepted by right-turning vehicles in the opposing flow.

Plotted values of the right-turning flow for values of α, B_r and B_0 typical of those noted when the opposing flow is in one or two lanes are shown in figure 39.1, which is reproduced from *Road Research Technical Paper* 56.

The difference between the average number of right-turning vehicles arriving per cycle and n_r calculated from equation 39.2 gives the number of vehicles that must turn right during an early cut-off or intergreen period.

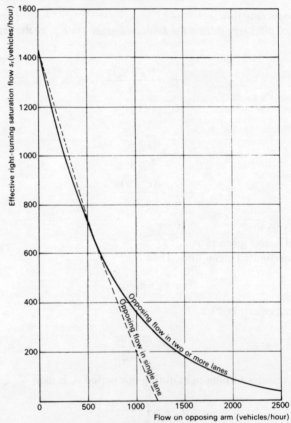

Figure 39.1 Right-turning saturation flows (based on *Road Research Technical Paper* 56)

	α	B_r	B_0
		(seconds)	
- - - Single-lane opposing flow	5	2.5	3
—— Two (or more) -lane opposing flow	6	2.5	1

α is the minimum gap required in the opposing flow for 1 right-turner
B_r is the minimum headway between successive right-turners ($1/B_r$ is the saturation flow)
B_0 is the minimum headway between successive vehicles in the opposing flow

Right-turning vehicles are assumed to discharge at headways of 2·5 s past a point in the centre of the intersection, the first right-turning vehicle passing this point at the end of the amber period and subsequent vehicles passing this point at intervals of 2·5 s.

If it is further assumed that the first vehicle on the cross phase needs 3 s from the start of the green indication on the cross phase to reach the mid-point of the intersection, then the length of the intergreen period will be 2·5 x the number of vehicles waiting to turn right.

The same procedure may be used to estimate the duration of any early cut-off period.

Problem

At an intersection with traffic flows as given in chapter 38 the effective green time on the west/east approach is 35 s, and the cycle time is 78 s. The right-turning saturation flow may be estimated from figure 39.1 for single-lane flow. The p.c.u./vehicle ratio is 1.2. Is the average number of vehicles waiting to turn right at the end of the green period approximately 2, 5 or 8?

Solution

Using equation 39.2

$$n_r = s_r \left(\frac{gs - qc}{s - q} \right)$$

Details of the traffic flows as given in chapter 38 and in this chapter are:

q (flow on the opposing east approach) = 700 p.c.u./h
or 700/1·2 = 583 vehicles/h

g (effective green time) = 35 s

c (cycle time) = 78 s

s (saturation flow on east approach) = 1900 p.c.u./h
or 1900/1·2 = 1583 vehicles/h

From figure 39.1, when q = 583 vehicles/h
s_r = 650 vehicles/h

then

$$n_r = \frac{650}{3600} \left(\frac{35 \times 1583 - 583 \times 78}{1583 - 583} \right)$$

$$= 1·8$$

Average number of right-turning p.c.u. arriving per cycle

$$= \text{flow} \times \text{cycle time}/3600$$

$$= \frac{200 \times 78}{3600}$$

$$= 4\cdot3 \text{ p.c.u.}$$

$$= 3\cdot6 \text{ vehicles}$$

The average number of vehicles waiting to turn right at the end of the green period is the average number of right-turning vehicles arriving per cycle minus the average numb able to turn right through the opposing flow per cycle; that is

$$3\cdot6 - 1\cdot8 = 1\cdot8 \text{ vehicles}$$

2 vehicles, approximately.

40

The ultimate capacity of
the whole intersection

It was explained in chapter 35 that the capacity of an approach is dependent on the lost time during the cycle. When the whole intersection is considered the capacity is also dependent on the total lost time on all the phases because the remainder of the time is shared equally between the phases and used as running time.

When the cycle time is calculated as described in chapter 36 and the green times apportioned as described in chapter 37, the approaches on each phase that have the highest ratio of flow to saturation flow (y value) will all reach capacity simultaneously as traffic growth takes place.

The ultimate capacity of the intersection is then the maximum flow that can pass through the intersection with the same relative flows and turning movements on the approaches.

As traffic growth takes place the capacity of the intersection can be increased by increasing the cycle time, because the ratio of lost time to cycle time decreases. There is however a practical limit beyond which there is little gain in efficiency as the cycle time is increased and at which drivers may become impatient at not receiving a green indication for the approach at which they are queueing. This practical limit is set at 120 s and the ultimate capacity of the intersection could be calculated as the flow that would just pass through the intersection when the signals were set at this cycle time.

This capacity is however the maximum capacity and is associated with long delays. It is more usual in traffic engineering work to use a practical capacity of 90 per cent of the maximum and this will result in shorter delays.

When traffic flows are uniform, the cycle time C_m, which is just long enough to pass all the traffic that arrives in one cycle, is easily calculated. Traffic flow is however semi-random or random and time is wasted because of variability of arrival times so that the minimum cycle time C_m is associated with extremely long delays.

The minimum cycle time is given by

$$C_m = L + \frac{q_1}{s_1} C_m + \frac{q_2}{s_2} C_m + \frac{q_3}{s_3} C_m + \cdots + \frac{q_n}{s_n} C_m$$

where q_n/s_n is the highest ratio of flow to saturation flow for phase n. Then

$$C_m = L + y_1C_m + y_2C_m + \cdots + y_nC_m$$
$$= L + C_m \times Y$$

or

$$Y = 1 - \frac{L}{C_m}$$

The maximum possible ratio of flow to saturation flow Y that can be accommodat over all the phases when C_m is at the practical limit and Y practical is employed is

$$Y_{\text{pract.}} = 0.9 \left(1 - \frac{L}{120}\right)$$

where $Y_{\text{pract.}}$ is 90 per cent of Y, or

$$Y_{\text{pract.}} = 0.9 - 0.0075L \tag{40.1}$$

As the Y value is the sum of the maximum ratios of flows divided by saturation flows for each phase and as the saturation flow is fixed, then as the flow increases Y increas from any given present day value of Y existing, to a future maximum value of Y practical. The reserve capacity at the existing flow is thus

$$\frac{100\,(Y_{\text{pract.}} - Y_{\text{exist.}})}{Y_{\text{exist.}}} \text{ per cent} \tag{40.2}$$

Problem

For the intersection with the flows given in chapter 38 and controlled by a two-phase system incorporating a late-start feature on the east approach, is the reserve capacity the intersection 4 per cent, 14 per cent or 24 per cent?

Solution .

The approaches and their y values are reproduced from chapter 38 in table 40.1.

TABLE 40.1

	Approach	Y value
1	west, straight ahead and left turning	0.21
2	west, right turning	0.13
3	east, all movements	0.37
4	north, all movements	0.26
5	south, all movements	0.32

The y values selected as y_{max} values are

east/west phase $0.13 + 0.37 = 0.50$
north/south phase $= 0.32$

From equation 40·1

$$Y_{\text{pract.}} = 0{\cdot}9 - 0{\cdot}0075L$$

and L = 6 s (see page 298)

$$Y_{\text{pract.}} = 0{\cdot}9 - 0{\cdot}0075 \times 6$$
$$= 0{\cdot}85$$

From equation 40.2

$$\text{reserve capacity} = \frac{100(0{\cdot}85 - 0{\cdot}82)}{0{\cdot}82}$$
$$= 4 \text{ per cent}$$

41

The optimisation of signal-approach dimensions

The procedures previously outlined have optimised cycle time and green times so as to produce minimum overall delay for an intersection where the physical dimensions of the approach highways were fixed.

When an intersection is being re-designed or a new intersection being constructed, then it is often desirable to modify the widths of the signal approaches. The necessity for this is not difficult to see, for if the intersection is to have the capacity of the appro highways, then it must have approximately twice the width of the highway if it has a green time of approximately half the real time.

Once the approach widths have been determined, then the optimum cycle time and the green times can be calculated for minimum delay conditions. Variations in approach width will result in differing optimum cycle times and thus there is theoretically an infinite number of variations possible. In practice the choice is less because the designe must normally produce a design with a standard lane width or multiples of this width.

One approach to this problem is that made by Webster and Newby[1]. They assumed that the maximum possible rate of flow across the stop line was proportional to the widths of the approach, w_1 and w_2, as shown in figure 41.1, and also that the widened sections of the approaches had lengths d_1 and d_2, which were just long enough to accommodate the queues that could pass through the intersection during fully saturate green periods.

For the intersection shown in figure 41.1 operating under two-phase control, q_1 and q_2 are the design flows of the first and second phase the approach widths of which are w_1 and w_2 and the effective green times of which are g_1 and g_2.

Then

$$cq_1 = sw_1g_1$$

and

$$cq_2 = sw_2g_2$$

where s is the saturation flow per unit width of road and c is the cycle time.

Since the sum of the effective green periods is constant

$$g_1 + g_2 = cq_1/sw_1 + cq_2/sw_2$$

314

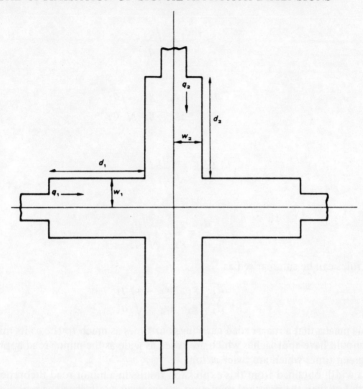

Figure 41.1 Idealised geometric layout for a signal-controlled junction with widened approaches

Differentiating with respect to w gives

$$\frac{dw_2}{dw_1} = \frac{-q_1}{q_2(w_1/w_2)^2}$$

It is required to minimise the total width W ($W = w_1 + w_2$) at the intersection. For minimum total width

$$\frac{dW}{dw_1} = 0$$

or

$$1 + \frac{dw_2}{dw_1} = 0$$

Substituting for dw_2/dw_1 gives

$$1 - \frac{q_1}{q_2(w_2/w_1)^2} = 0$$

or

$$\frac{w_1}{w_2} = \sqrt{\frac{q_1}{q_2}}$$

and

$$\frac{q_1}{q_2} = \frac{w_1 g_1}{w_2 g_2} = \frac{g_1}{g_2}\sqrt{\frac{q_1}{q_2}}$$

or

$$\frac{g_1}{g_2} = \sqrt{\frac{q_1}{q_2}}$$

Also

$$\frac{d_1}{d_2} = \frac{g_1}{g_2} = \sqrt{\frac{q_1}{q_2}}$$

these rules can be summarised as

$$\frac{d_1}{d_2} = \frac{g_1}{g_2} = \frac{w_1}{w_2} = \sqrt{\frac{q_1}{q_2}} \qquad (41.1)$$

This means that a major road carrying four times as much traffic as its minor cross-road should have approaches which are twice as wide as the minor road approaches and have green times which are twice as long.

If a width obtained from this expression results in a minor road theoretically having a width less than the practical minimum carriageway width then the practical minimum should be employed and a green time less than that theoretically necessary used. The green time saved can then be allocated to the other phases so allowing a reduction in their width.

It is often found that the flows on the differing arms of the same phase are approximately equal even though they may occur at differing times of the day. If this is not so then the highest flow should be used in the formula to obtain the width and the green time. Next the width of the approach on the same phase with the lower flow can be obtained.

The same rule can be applied to multi-phase intersections, when

$$w_1 : w_2 \ldots w_n$$
$$g_1 : g_2 \ldots g_n$$
$$d_1 : d_2 \ldots d_n$$

are proportional to

$$\sqrt{q_1} : \sqrt{q_2} \ldots \sqrt{q_n}$$

With T-junctions with 2-phase control the ratios of widths, green times and widened lengths should be

$$\frac{w_1}{w_2} = \sqrt{\frac{q_1}{2q_2}}$$

and

$$\frac{g_1}{g_2} = \frac{d_1}{d_2} = \sqrt{\frac{2q_1}{q_2}}$$ (41.2)

where q_1 etc., refers to the major road,
q_2 etc., refers to the minor road.

Reference

1. F. V. Webster and R. F. Newby. Research into the relative merits of roundabouts and traffic-signal controlled intersections. *J. Instn civil Engrs*, **27** (Jan 1964), 47–76

Problem

The traffic flows at an intersection are given in table 41.1

TABLE 41.1

Approach	Flow (p.c.u./h)
north	3600
south	3300
east	800
west	900

Show that the relative proportions of the approach widths and effective green times on the widest approaches of each phase are 2:1 when two-phase control is used and the design minimises the area of land required.

Show also that when:

(a) the cycle time is 120 s;
(b) the total lost time per cycle is 6 s;
(c) the intergreen period is 4 s;
(d) the minimum approach width on the east/west approach gives a saturation flow of 3300 p.c.u./h;

the actual green settings are 32 s and 80 s.

Solution

The two phases used for control purposes will be the north/south flow and the east/west flow (see page 298).

Select the maximum flow on each phase

north   3600 p.c.u./h
west   900 p.c.u./h

From equation 41.1

$$\frac{g_1}{g_2} = \frac{w_1}{w_2} = \sqrt{\frac{3600}{900}} = 2$$

The relative proportions of widths and effective green times for the widest approach on each phase is thus 2:1.

When the cycle time is 120 s and the total lost time per cycle is 6 s, then the available effective green time is (see page 298)

$$120 - 6 = 114 \text{ s}$$

When this is divided between the phases in the ratio of 2:1 then

effective green time, north/south phase = 76 s

effective green time, east/west phase = 38 s

For the west approach the minimum width is stated to give a saturation flow of 3300 p.c.u./h.

The maximum flow that can be passed with the signal settings as calculated is

$$\frac{\text{effective green time}}{\text{cycle time}} \times 3300 \text{ p.c.u./h} = \frac{38 \times 3300}{120}$$

$$= 1045 \text{ p.c.u./h}$$

This maximum flow is greater than the design flow on the west approach, which is given as 900 p.c.u./h. It is therefore possible to reduce the green time from the value calculated using equation 41.1. The effective green time required can be calculated from

$$\text{actual flow} = \frac{\text{saturation flow} \times \text{effective green time}}{\text{cycle time}}$$

or

$$\text{effective green time} = \frac{900 \times 120}{3300} = 33 \text{ s (approx.)}$$

The effective green time as calculated from equation 41.1 is 38 s and this leaves 38 − 33 = 5 s available for use on the north/south phase. The total effective green time on the north/south phase is then 76 + 5 = 81 s.

As the intergreen period is 4 s and the total lost time per cycle is 6 s then starting delays are 2 s of each green period (see pages 291 and 298).

actual green time = effective green time + lost time due to starting delays

− amber period

= effective green time − 1

= 80 s and 32 s

42

Optimum signal settings when saturation flow falls during the green period

On some traffic-signal approaches it cannot be assumed that saturation flow will remain constant throughout the greater part of the green period. The effect of blocked right-turning movements and of approaches that are wider at the stop line than on the remainder of the approach is to reduce the flow rate over the stop line as the green period proceeds.

In this problem it is the saturation flow just as the amber period commences that is important because it is this value that determines the variation in the number of

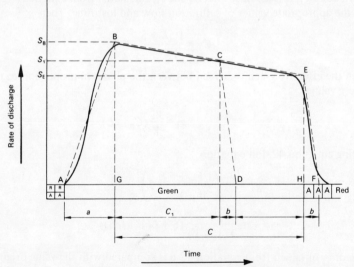

Figure 42.1 Variation of discharge across a traffic-signal stop line where saturation flow decreases during the green period (based on *Road Research Technical Paper* 56)

vehicles crossing the stop line under saturated conditions when there is a small change in green time. It is thus necessary to find the length of green time that corresponds to the saturation flow just as amber begins and that, when substituted into the formula for optimum setting, produces the original green time.

The change in flow rate with time is illustrated for a generalised case in figure 42.1.

The rate of discharge against time curve may be replaced by a quadrilateral with the same area; this is shown by dashed lines. Then the areas ABG and EHF are equal to the areas under the corresponding portions of the curves. The saturation flow at B is S_B and at E is S_E. If the green time ends earlier at C then the saturation flow at the commencement of amber is S_1 and the reduction in flow during the amber period can be denoted by the effective line CD. From figure 42.1

$$S_1 = S_B - \frac{C_1}{C}(S_B - S_E) \tag{42.1}$$

The effective green time g_1 when the green time terminates at C can be calculated from the area of the quadrilateral ABCD, which is $\frac{1}{2}S_B a + \frac{1}{2}(S_B + S_1)C_1 + \frac{1}{2}S_1 b$. From the original definition of effective green time this area is equal to $g_1 S_1$.

Hence

$$g_1 = \frac{S_B(a + C_1) + S_1(b + C_1)}{2S_1} \tag{42.2}$$

If G is the combined green plus amber period, the lost time is

$$l_1 = G - g_1 \tag{42.3}$$

The value of g_1 is not necessarily the optimum green time. To obtain a first approximation to the optimum green time, which can then be compared with the originally assumed value of g_1, it is assumed that S_1 and l_1, calculated from equations 42.1 and 42.3, are the appropriate values of saturation flow and lost time. Then

$$C_0 = \frac{1 \cdot 5L + 5}{1 - Y} \tag{42.4}$$

Also when the effective green is divided between the phases in proportion to their respective y values

$$g_1 = \frac{y_1}{Y}(C_0 - L) \tag{42.5}$$

Substituting equation 42.4 in equation 42.5

$$g_1 = \frac{y_1}{Y}\left(\frac{1 \cdot 5L + 5}{1 - Y} - L\right)$$

$$= \frac{y_1}{Y(1 - Y)}(5 + L(Y + 0 \cdot 5)) \tag{42.6}$$

The value of g_1 obtained from equation 42.6 is compared with the value originally obtained from equation 42.2. A new estimate of g_1 may then be made by taking a second approximation midway between the previous two values. The iteration may then be repeated until there is no significant difference between successive values of g_1. This method is described in detail in Appendix 6 of reference 2 in chapter 29.

Problem

The discharge/time curve at a traffic-signal stop line for the critical approach of phase (a) may be approximated by the curve shown in figure 42.2.

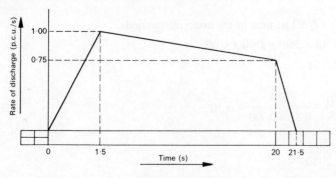

Figure 42.2

On the other approaches the saturation flow is uniform with time. The design hour flows and saturation flows for the approaches on each phase with the maximum y values are given in table 42.1.

TABLE 42.1

Approach	Design hour flow (p.c.u./h)	Saturation flow (p.c.u./h)
phase 1	600	shown graphically
phase 2	500	2,000

Intergreen periods 4 s, lost time due to starting delays on phase 2 is 2 s.
   Calculate the optimum cycle time and the corresponding actual green times.

Solution

The iterative procedure to arrive at the optimum cycle time will be illustrated by carrying out the calculation in steps.

1. Assume the actual green plus amber period on phase 1 is 18 s (point C, figure 42.1).

2. $S_1 = 1 \cdot 0 - \dfrac{13 \cdot 5}{18 \cdot 5}(1 \cdot 00 - 0 \cdot 75)$

   $= 0 \cdot 82$ p.c.u./s   (see equation 42.1).

3. $g_1 = \dfrac{1 \cdot 0(1 \cdot 5 + 13 \cdot 5) + 0 \cdot 82(1 \cdot 3 + 13 \cdot 5)}{2 \times 0 \cdot 82}$

     $= 16 \cdot 55$ s   (see equation 42.2).

4. $l_1 = 18 - 16 \cdot 55$

    $= 1 \cdot 45$ s   (see equation 42.3).

5. $L = l_1 + l_2 +$ lost time in the intergreen periods

    $= 1 \cdot 45 + 2 \cdot 00 + 1 \cdot 00 + 1 \cdot 00$

    $= 5 \cdot 45$ s

6. $y_1 = \dfrac{q_1}{S_1} = \dfrac{600}{0 \cdot 82 \times 3600} = 0 \cdot 30$

   $y_2 = \dfrac{500}{2000} = 0 \cdot 25.$

7. $g_1 = \dfrac{0 \cdot 30}{0 \cdot 55(1 - 0 \cdot 55)} \, 5 + 5 \cdot 45(0 \cdot 55 + 0 \cdot 5)$

    $= 12 \cdot 86$ s   (see equation 42.6)

8. The first calculation of actual green plus amber period on phase 1 is

        $12 \cdot 86$ s $+ 1 \cdot 45$ s $= 14 \cdot 31$ s   (equation 42.3)

The initial assumption of actual green plus amber period produced a value of $g_1 = 16 \cdot 55$ s in Step 3 and finally in Step 7 it was calculated as $12 \cdot 86$ s.

A second iteration may now be performed using as the initial assumption that the actual green plus amber period is

$$18 - \frac{16 \cdot 55 - 12 \cdot 86}{2} = 16 \cdot 06 \text{ s}.$$

Successive iterations are tabulated in table 42.2.

    effective green time phase $1 = 13 \cdot 52$ s

    effective green time phase $2 = \dfrac{13 \cdot 52 \times 0 \cdot 25}{0 \cdot 30}$

          $= 11 \cdot 27$ s   (see chapter 37)

    actual green time phase $1 = 13 \cdot 52 + 1 \cdot 86 - 3 \cdot 00 = 12 \cdot 38$ s

    actual green time phase $2 = 11 \cdot 27 - 1 \cdot 00 = 10 \cdot 27$ s   (see chapter 35)

      cycle time $= 12 \cdot 38 + 10 \cdot 27 + 8 \cdot 00 = 30 \cdot 65$ s   (see chapter 37)

TABLE 42.2

Step	Description	Second iteration	Third iteration
1	initial value of green plus amber period	16·06 s	15·72
2	$S_1 = 1·0 - \dfrac{11·56}{18·5}(1·00 - 0·75)$	0·84 p.c.u./s	
3	$g_1 = \dfrac{1·0(1·5 + 11·56) + 0·84(1·3 + 11·56)}{1·68}$	14·20 s	the values of the green plus amber period in the second and third iterations are sufficiently close to make further iterations unnecessary
4	$l_1 = 16·06 - 14·20$	1·86 s	
5	$L = l_1 + l_2 +$ intergreen lost time	5·86 s	
6	$y_1 = \dfrac{600}{0·84 \times 3600}$	0·30	
	$y_2 = \dfrac{500}{2000}$	0·25	
7	$g_1 = \dfrac{0·30}{0·55(1 - 0·55)}(5 + 5·86(0·55 + 0·5))$	13·52 s	

43

Delay at signal-controlled intersections

As has been briefly mentioned in chapter 36, the delay to vehicles on a traffic-signal approach has been investigated by F. V. Webster and reported in *Road Research Technical Paper* 39. By a combination of queueing theory and digital computer simulation it has been shown that the average delay per vehicle on a particular intersection arm is given by

$$d = \frac{c(1 - \lambda)^2}{2(1 - \lambda x)} + \frac{x^2}{2q(1 - x)} - 0 \cdot 65 \left(\frac{c}{q^2}\right)^{1/3} x^{(2 + 5\lambda)}$$

where d = average delay per vehicle,

c = cycle time,

λ = proportion of the cycle that is effectively green for the phase under consideration (that is, effective green time/cycle time),

q = flow,

s = saturation flow,

x = degree of saturation, which is the ratio of actual flow to the maximum flow that can be passed through the approach (that is $q/\lambda s$).

The relative value of the three terms is illustrated in figure 43.1 reproduced from *Road Research Technical Paper* 39. The first term in this expression is the delay due to a uniform rate of vehicle arrival, the second term is the delay due to the random nature of the vehicle arrivals. The third term was empirically derived from the simulation of traffic flow.

To assist in the calculation of delays on traffic signal approaches values of

$$\frac{(1 - \lambda)^2}{2(1 - \lambda x)} = A$$

$$\frac{x^2}{2(1 - x)} = B$$

correction term = C

are given in *Road Research Technical Paper* 39 and are tabulated in tables 43.1–3. It is

324

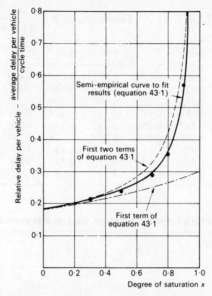

Figure 43.1 Illustration of delay on traffic-signal approaches (based on *Road Research Technical Paper 39*)

hen possible to calculate the delay from

$$d = cA + \frac{B}{q} - C \qquad (43.1)$$

TABLE 43.1 Tabulation of $A = \dfrac{(1 - \lambda)^2}{2(1 - \lambda x)}$

λ x	0·1	0·2	0·3	0·35	0·40	0·45	0·50	0·55	0·60	0·65	0·70	0·80	0·90
0·1	0·409	0·327	0·253	0·219	0·188	0·158	0·132	0·107	0·085	0·066	0·048	0·022	0·005
0·2	0·413	0·333	0·261	0·227	0·196	0·166	0·139	0·114	0·091	0·070	0·052	0·024	0·006
0·3	0·418	0·340	0·269	0·236	0·205	0·175	0·147	0·121	0·098	0·076	0·057	0·026	0·007
0·4	0·422	0·348	0·278	0·246	0·214	0·184	0·156	0·130	0·105	0·083	0·063	0·029	0·008
0·5	0·426	0·356	0·288	0·256	0·225	0·195	0·167	0·140	0·114	0·091	0·069	0·033	0·009
0·55	0·429	0·360	0·293	0·262	0·231	0·201	0·172	0·145	0·119	0·095	0·073	0·036	0·010
0·60	0·431	0·364	0·299	0·267	0·237	0·207	0·179	0·151	0·125	0·100	0·078	0·038	0·011
0·65	0·433	0·368	0·304	0·273	0·243	0·214	0·185	0·158	0·131	0·106	0·083	0·042	0·012
0·70	0·435	0·372	0·310	0·280	0·250	0·221	0·192	0·165	0·138	0·112	0·088	0·045	0·014
0·75	0·438	0·376	0·316	0·286	0·257	0·228	0·200	0·172	0·145	0·120	0·095	0·050	0·015
0·80	0·440	0·381	0·322	0·293	0·265	0·236	0·208	0·181	0·154	0·128	0·102	0·056	0·018
0·85	0·443	0·386	0·329	0·301	0·273	0·245	0·217	0·190	0·163	0·137	0·111	0·063	0·021
0·90	0·445	0·390	0·336	0·308	0·281	0·254	0·227	0·200	0·174	0·148	0·122	0·071	0·026
0·92	0·446	0·392	0·338	0·312	0·285	0·258	0·231	0·205	0·179	0·152	0·126	0·076	0·029
0·94	0·447	0·394	0·341	0·315	0·288	0·262	0·236	0·210	0·183	0·157	0·132	0·081	0·032
0·96	0·448	0·396	0·344	0·318	0·292	0·266	0·240	0·215	0·189	0·163	0·137	0·086	0·037
0·98	0·449	0·398	0·347	0·322	0·296	0·271	0·245	0·220	0·194	0·169	0·143	0·093	0·042

TABLE 43.2   Tabulation of $B = \dfrac{x^2}{2(1-x)}$

x	0·00	0·01	0·02	0·03	0·04	0·05	0·06	0·07	0·08	0·09
0·1	0·006	0·007	0·008	0·010	0·011	0·013	0·015	0·017	0·020	0·022
0·2	0·025	0·028	0·031	0·034	0·038	0·042	0·046	0·050	0·054	0·055
0·3	0·064	0·070	0·075	0·081	0·088	0·094	0·101	0·109	0·116	0·125
0·4	0·133	0·142	0·152	0·162	0·173	0·184	0·196	0·208	0·222	0·235
0·5	0·250	0·265	0·282	0·299	0·317	0·336	0·356	0·378	0·400	0·425
0·6	0·450	0·477	0·506	0·536	0·569	0·604	0·641	0·680	0·723	0·768
0·7	0·817	0·869	0·926	0·987	1·05	1·13	1·20	1·29	1·38	1·49
0·8	1·60	1·73	1·87	2·03	2·21	2·41	2·64	2·91	3·23	3·60
0·9	4·05	4·60	5·28	6·18	7·36	9·03	11·5	15·7	24·0	49·0

TABLE 43.3   Correction term of equation 43.1 as a percentage of the first two terms

x	λ \ M*	2·5	5	10	20	40
0·3	0·2	2	2	1	1	0
	0·4	2	1	1	0	0
	0·6	0	0	0	0	0
	0·8	0	0	0	0	0
0·4	0·2	6	4	3	2	1
	0·4	3	2	2	1	1
	0·6	2	2	1	1	0
	0·8	2	1	1	1	1
0·5	0·2	10	7	5	3	2
	0·4	6	5	4	2	1
	0·6	6	4	3	2	2
	0·8	3	4	3	3	2
0·6	0·2	14	11	8	5	3
	0·4	11	9	7	4	3
	0·6	9	8	6	5	3
	0·8	7	8	8	7	5
0·7	0·2	18	14	11	7	5
	0·4	15	13	10	7	5
	0·6	13	12	10	8	6
	0·8	11	12	13	12	10
0·8	0·2	18	17	13	10	7
	0·4	16	15	13	10	8
	0·6	15	15	14	12	9
	0·8	14	15	17	17	15
0·9	0·2	13	14	13	11	8
	0·4	12	13	13	11	9
	0·6	12	13	14	14	12
	0·8	13	13	16	17	17
0·95	0·2	8	9	9	9	8
	0·4	7	9	9	10	9
	0·6	7	9	10	11	10
	0·8	7	9	10	12	13

	0·2	8	9	10	9	8
0·975	0·4	8	9	10	10	9
	0·6	8	9	11	12	11
	0·8	8	10	12	13	14

* M is the average flow per cycle = qc.

Problems

An intersection controlled by two-phase traffic signals has the design-hour flows and saturation flows tabulated in table 43.4.

TABLE 43.4

Approach	Flow (vehicles/h)	Actual green (s)	Saturation flow (vehicles/h)
north	2000	80	3350
south	1750	80	3350
east	660	32	2750
west	750	32	2750

Starting delays 2 s of each green plus amber period intergreen period 4 s; cycle time 120 s.

For traffic flow on the north approach, is the delay element due to the random arrival of vehicles greater than the element due to the regular arrival of vehicles?

For the traffic flow on the east approach, is the average delay per vehicle approximately 3·0 s, 60 s or 9·0 s?

Solutions

First calculate the parameters of equation 43.1. For the north approach

$$c = 120 \text{ s}$$

$$\lambda = \text{effective green/cycle time}$$

$$= (80 + 1)/120 \quad \text{(see page 291)}$$

$$= 0.675$$

$$q = 2000/3600$$

$$= 0.556 \text{ vehicles/s}$$

$$s = 3350/3600$$

$$= 0.931 \text{ vehicles/s}$$

$$x = q/\lambda s$$

$$= 0.556/(0.675 \times 0.931)$$

$$= 0.885$$

The delay element due to the uniform rate of arrival of vehicles is given by the first term of equation 43.1, that is cA.

Tabulated values of A are given in table 43.1 and referring to this table and using th approximate values $\lambda = 0 \cdot 70$, $x = 0 \cdot 90$

$$A = 0 \cdot 122$$

and

$$d = 120 \times 0 \cdot 122 = 14 \cdot 64 \text{ s}$$

The delay element due to the random rate of arrival of vehicles is given by the seco term of equation 43.1, that is B/q. Tabulated values of B are given in table 43.2. Referring to this table and using the approximate values for x of $0 \cdot 89$

$$B = 3 \cdot 60$$

Then

$$d = 3 \cdot 60/0 \cdot 556 \text{ s}$$
$$= 6 \cdot 47 \text{ s}$$

Note. The delay element due to the uniform arrival of vehicles ($14 \cdot 64$ s) is greater tha the delay element due to the random arrival of vehicles ($6 \cdot 47$ s).

First calculate the parameters of equation 43.1 for the east approach

$$c = 120 \text{ s}$$
$$\lambda = (32 + 1)/120 \quad \text{(see page 291)}$$
$$= 0 \cdot 275$$
$$q = 660/3600$$
$$= 0 \cdot 183 \text{ vehicle/s}$$
$$s = 2750/3600$$
$$= 0 \cdot 764 \text{ vehicle/s}$$
$$x = 0 \cdot 183/(0 \cdot 275 \times 0 \cdot 764)$$
$$= 0 \cdot 871$$
$$M = qc$$
$$= 0 \cdot 183 \times 120$$
$$= 21 \cdot 96 \text{ vehicles/cycle}$$

From equation 43.1

$$d = cA + \frac{B}{q} - C$$

From table 43.1, $A = 0 \cdot 329$.
From table 43.2, $B = 2 \cdot 91$.
From table 43.3, $C = 11$ per cent of first two terms
and

$$d = 49 \cdot 3 \text{ s/vehicle}$$

44

Determination of the optimum cycle from a consideration of delays on the approach

Minimisation of delay for a complete intersection has been investigated in Appendix 3, *Road Research Technical Paper* 39 by considering the first two terms in equation 43.1. The total delay for the intersection is given by

$$D = \sum (\text{average delay per vehicle}) \times \text{flow}$$

$$D = \sum_{r=1}^{r=n} \left[\frac{c(1 - \lambda_r)^2}{2(1 - \lambda_r x_r)} + \frac{x_r^2}{2q_r(1 - x_r)} \right] q_r$$

$$= \sum_{1}^{n} \left[\frac{c y_r s_r (1 - \lambda_r)^2}{2(1 - y_r)} + \frac{y_r^2}{2q_r(\lambda_r - y_r)} \right]$$

where $y = q/s$ for any phase.

Differentiating with respect to cycle time

$$\frac{dD}{dc} = \sum_{1}^{n} \left[\frac{(1 - \lambda_r)^2 y_r s_r}{2(1 - y_r)} - \frac{d\lambda_r}{dc} \frac{y_r^2(2\lambda_r - y_r)}{2\lambda_r^2(\lambda_r - y_r)} + \frac{c y_r s_r (1 - \lambda_r)}{(1 - y_r)} \right]$$

$$= 0 \text{ for minimum delay}$$

that is

$$\sum_{1}^{n} \frac{y_r s_r (1 - \lambda_r)}{1 - y_r} \left(\frac{1 - \lambda_r}{2} - C_0 \frac{d\lambda_r}{dc} \right) - \sum_{1}^{n} \frac{y_r^2(2\lambda_r - y_r)}{2\lambda_r^2(\lambda_r - y_r)^2} \cdot \frac{d\lambda_r}{dc} = 0$$

329

If λ is made proportional to y, then

$$\lambda_r = \frac{c - L}{c} \times \frac{y_r}{Y}$$

and

$$\frac{d\lambda_r}{dc} = \frac{y_r L}{Y c^2}$$

Substituting these values gives

$$\frac{1}{Y} \sum_1^n \frac{y_r s_r}{1 - y_r} (C_0^2(Y - y_r)^2 - L^2 y_r^2) - L \sum_1^n \frac{y_r^2(2\lambda_r - y_r)}{\lambda_r^2(\lambda_r - y_r)^2} = 0$$

and this further reduces to

$$\frac{1}{Y} \sum_1^n \frac{y_r s_r}{1 - y_r} [C_0^2(Y - y_r)^2 - L^2 y_r^2][(C_0^2 - 2C_0 L + L^2)]$$
$$- LY^3 C_0^3 n \times \frac{[C_0(2 - Y) - 2L]}{[C_0(1 - Y) - L]^2} = 0 \quad (4$$

which can be further simplified by approximating the terms $L^2 y_r^2$ and L^2, which ar
small in comparison with the other terms.

If values of average delay are calculated using equation 43.1 then a family of cur
as shown in figure 36.1 are obtained. From an inspection of these curves it can be s
that the minimum cycle is approximately equal to twice the optimum cycle, which
represented by the vertical asymptote to the curve. It is thus possible to obtain a fir
approximation by this relationship.

The minimum cycle is just long enough to pass the traffic that arrives in one cycl
it is associated with very high delays. The minimum cycle is equal to the lost time p
cycle and the amount of time necessary to pass all the traffic through the intersecti
at the maximum possible rate; that is

$$C_m = L + \frac{q_1}{s_1} C_m + \frac{q_2}{s_2} C_m + \cdots + \frac{q_n}{s_n} C_m$$
$$= L + C_m(y_1 + y_2 + \cdots + y_n)$$
$$= L + C_m Y$$

that is

$$C_m = \frac{L}{1 - Y}$$

or

$$C_0 = \frac{2L}{1 - Y}$$

It is now possible to replace L by $C_0(1 - Y)/2$ in those terms which in equation
are comparatively small, that is $L^2 y_r^2$ and L^2.

The term

$$[C_0^2(Y - y_r)^2 - L^2 y_r^2]$$

becomes

$$C_0^2 \left[(Y - y_r)^2 - \frac{y_r^2}{4}(1 - Y)^2 \right]$$

and the term

$$[(C_0^2 - 2C_0 L) + L^2]$$

becomes

$$(C_0^2 - 2C_0 L) \left[1 + \frac{L^2}{C_0^2 - 2C_0 L} \right] = (C_0^2 - 2C_0 L) \left[1 + \frac{(L/C_0)^2}{(1 - 2L/C_0)} \right]$$

$$= C_0(C_0 - 2L) \frac{(1 + Y)^2}{4Y}$$

Substituting these terms in equation 44.1 gives

$$\frac{(1 + Y)^2}{16Y^5 n} \times L(C_0 - 2L) \sum_{1}^{n} \frac{y_r s_r}{1 - y_r} [4(Y - y_r)^2 - y_r^2(1 - Y)^2]$$

$$- \frac{L^2[C_0(2 - Y) - 2L]}{[C_0(1 - Y) - L]^2} = 0$$

Let

$$E = \frac{L(1 + Y)^2}{16n Y^5} \sum_{1}^{n} \frac{y_r s_r}{1 - y_r} [4(Y - y_r)^2 - y_r^2(1 - Y)^2]$$

Then the equation reduces to

$$(C_0 - 2L)E - \frac{L^2[C_0(2 - Y) - 2L]}{[C_0(1 - Y) - L]^2} = 0$$

or

$$C_0^3(1 - Y)^2 - 2C_0^2 L(1 - Y)(2 - Y) + C_0 L^2 \left[5 - 4Y - \frac{(2 - Y)}{E} \right]$$

$$+ L^3 \left[\frac{2}{E} - 2 \right] = 0 \quad (44.2)$$

C_0 has been taken to be approximately $2L/(1 - Y)$; or a more accurate value can be taken as $2L/(1 - Y)F$ where F is a factor that takes into account the flows, the saturation flows and the lost time at the intersection. This makes it possible to replace L^3 in the last term in equation 44.2 to give

$$L^2 \frac{C_0(1 - Y)}{2F}$$

It is then possible to divide throughout the equation by C_0 and solve the resulting quadratic equation. This gives

$$C_0 = \frac{2L}{1-Y} \left\{ 1 + \frac{\sqrt{[Y^2 - Y/F + 1/E + (1/F - 1)(1 - (1 - Y)/E)] - Y}}{2} \right\} \quad (44.3)$$

For most purposes P is close enough to unity for equation 44.3 to be rewritten as

$$C_0 = \frac{2L}{1-Y} \left\{ 1 + \frac{\sqrt{[Y^2 - Y + 1/E] - Y}}{2} \right\} \quad (44.4)$$

Also

$$C_0 = \frac{2L}{1-Y} F$$

so that for a first approximation

$$F = \left\{ 1 + \frac{\sqrt{[Y^2 - Y + 1/E] - Y}}{2} \right\} \quad (44.5)$$

When evaluating C_0 the first step is to calculate F from equation 44.5. If its value is not more than 10 per cent from unity then C_0 may be calculated from

$$C_0 = \frac{2LF}{1-Y}$$

If however F is not close to unity then equation 44.3 should be used for a more accurate calculation.

This procedure is rather complicated for practical use and a relationship has been derived between $2LF$ and an expression of the form $KL + 5$. The value of K varies with junction type and with flow ratios but the most usual value adopted in practice is $K = 1 \cdot 5$ giving the well-known formula

$$C_0 = \frac{1 \cdot 5L + 5}{1 - Y} \qquad \text{(see page 294)}$$

The evaluation of C_0 from this relationship is tedious and a procedure to simplify the calculation is given in *Road Research Technical Paper* 39.

The procedure is as follows:

1. For each phase calculate $G = (3 - Y)/2Y$ and multiply each y value by G giving Gy_r for each phase.
2. From figure 44.1 determine B_r for each phase and multiply by the appropriate value of the saturation flow s_r to give $\Sigma s_r B_r$ for all n phases.

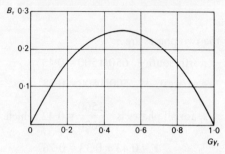

Figure 44.1 Graphical determination of B_r (based on *Road Research Technical Paper* 39)

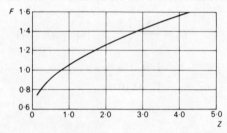

Figure 44.2 Graphical determination of F (based on *Road Research Technical Paper* 39)

3. Calculate $Z = n/L \sum\limits^{n} s_r B_r$.
4. From figure 44.2 determine F.
5. Calculate C_0 from

$$C_0 = \frac{2LF}{1 - Y}$$

Problems

Two-phase traffic-signal control is used to resolve conflicts at an intersection with the following flows and saturation flows (table 44.1). The total lost time per cycle is 6 s and the intergreen period is 4 s.

TABLE 44.1

Approach	Design-hour flow (vehicles/h)	Saturation flow (vehicles/h)
north	550	1500
south	650	1500
east	450	1500
west	500	1500

Show that the error involved in using the approximate practical formula for the optimum cycle time rather than the more accurate procedure described in *Road Research Technical Paper* 39 is less than 5 per cent.

Solutions

The y_{max} values for the two phases are

$$\text{north/south} \quad 650/1500 = 0\cdot43$$

$$\text{east/west} \quad 500/1500 = 0\cdot33$$

$$s_r \text{ for all phases is} \quad \frac{1500}{3600} = 0\cdot42 \text{ vehicle/s}$$

$$Y = 0\cdot43 + 0\cdot33 = 0\cdot76$$

The approximate practical formula for the optimum cycle time is given by

$$C_0 = \frac{1\cdot5L + 5}{1 - Y} \quad \text{(see page 294)}$$

$$= \frac{1\cdot5L + 5}{1 - 0\cdot76}$$

$$= 58\cdot3 \text{ s}$$

The more accurate procedure is as follows:

1. For the north/south phase

$$Gy_r = y \times \frac{3 - Y}{2Y} = 0\cdot43 \times \frac{3 - 0\cdot76}{2 \times 0\cdot76}$$

$$= 0\cdot63$$

For the east/west phase

$$Gy_r = y \times \frac{3 - Y}{2Y} = 0\cdot33 \times \frac{3 - 0\cdot76}{2 \times 0\cdot76}$$

$$= 0\cdot44$$

2. From figure 44.1

For the north/south phase and east/west phase

$$B_r = 0\cdot24$$

$$s_r B_r = 0\cdot42 \times 0\cdot24$$

$$= 0\cdot10$$

3. $Z = n/L \sum\limits_{}^{n} s_r B_r$

$$= 2/6(0\cdot10 + 0\cdot10)$$

$$= 1\cdot67$$

4. From figure 44.2

$$F = 1 \cdot 2$$

5. $C_0 = \dfrac{2L}{1 - Y} \times F$

$\quad = \dfrac{2 \times 6 \times 1 \cdot 2}{1 - 0 \cdot 76}$

$\quad = 60 \cdot 0 \text{ s}$

The error in the approximate practical formula is then

$$\left(\frac{60 \cdot 0 - 58 \cdot 3}{60} \right) 100 \text{ per cent}$$

that is 2·8 per cent, which is less than 5·0 per cent.

45

Average queue lengths at the commencement of the green period

In the design of traffic signals it is often desirable to be able to estimate the queue length at the beginning of the green period. The queue length at this period of the cyc will normally be the greatest experienced because during the green period the queue is being discharged at a rate which for practical purposes must be greater than the flow on the approach.

Of the several cases that may be considered, the simplest one is the unsaturated approach where the queue at the beginning of the green period disappears before the signal aspect changes to amber. In this case the maximum queue is simply the number of vehicles that have arrived during the preceding effective red period, that is

$$N_u = qr \tag{45.1}$$

where N_u is the initial queue at the beginning of an unsaturated green period,
q is the flow,
r is the length of the effective red period; the effective red period is equal to the cycle time minus the effective green period.

The next simplest case is when the approach is fully saturated. The queue length wi have gradual variations over the interval considered, on which are superimposed sudde increases and decreases caused by the red and green periods. The range of the short variations will be qr and the average queue over the whole of the interval considered (assuming no great variation between beginning and end) is equal to the product of th flow and the average delay per vehicle.

The average queue at the beginning of the green period is thus equal to the average queue throughout the interval plus half the average range of the cyclic fluctuations; th is

$$N_s = qd + \tfrac{1}{2}qr$$

336

where N_s is the initial queue at the beginning of a saturated green period.

d is the average delay per vehicle on a single approach.

Note that when the approach is only just saturated then d is $qr/2$ and $N_s = qr$, the same as for an unsaturated approach. The value of N_s cannot be less than qr so that

$$N_s = qd + \tfrac{1}{2}qr \qquad \text{or} \qquad qr, \qquad (45.2)$$

whichever is the greater.

In most practical cases intersections will operate under a mixture of saturated and unsaturated conditions. The initial queue at the commencement of the green period for these conditions will now be discussed.

Assume there are n fully saturated cycles and m unsaturated cycles. The flow is q and the average delay per vehicle d over $(m + n)$ cycles. These are composed of flow q_s and delay d_s for n cycles and flow q_u with delay d_u for m cycles.

From the previous discussion the average queue at the beginning of the green period for the n saturated cycles will be

$$\frac{q_s r}{2} + q_s d_s$$

Similarly for the m unsaturated cycles. The average queue at the beginning of the green period for the m unsaturated cycles will be

$$q_u r$$

Therefore the average queue at the beginning of the green period over $(m + n)$ cycles will be

$$N = \frac{1}{(m + n)}\left(nq_s r/2 + nq_s d_s + mq_u r\right)$$

The average flow is given by

$$q = \frac{nq_s + mq_u}{(n + m)}$$

so that

$$N = qr/2 + \frac{nq_s d_s + mq_u r/2}{n + m}$$

The second expression on the right-hand side presents difficulties in evaluating a solution so it must be converted into a more easily manipulated form.

Considering the unsaturated cycles, the net rate of discharge of the queue is $s - q_u$, the average queue at the beginning of the green period is N and the time taken for this initial average queue to disperse is $N/(s - q_u)$. The average queue throughout the cycle is

$$\frac{N}{2c}\left(\frac{N}{s - q_u} + r\right)$$

as illustrated in figure 45.1.

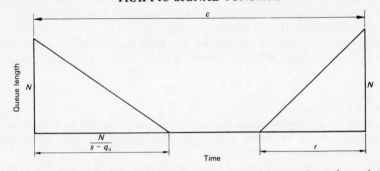

Figure 45.1 Variation of queue with time when the approach is operating under a mixture of saturated and unsaturated conditions

Average delay = average queue × cycle time/number of vehicles per cycle.

$$d_u = \text{average queue}/q_u$$

$$d_u = \frac{r}{2c}\left(\frac{rs}{c(s - q_u)}\right)$$

This is approximately $r/2$ an approximation that underestimates delay. But d_u is very much smaller than d_s and m is also much smaller than n, so that

$$N = \frac{qr}{2} + \frac{nq_s d_s + mq_u d_u}{n + m}$$

$$N = \frac{qr}{2} + qd$$

We know also that the average queue at the beginning of the green period cannot be le than qr so that the value of N is usually taken as the greater of these two values.

This theoretical approach neglects the finite extent of the queue and in practice vehicles join the queue earlier and so have an increased delay. This increase can be calculated from

length of queue when the green period begins = Nj/a

where a = number of lanes in queue,

$\quad j$ = average spacing of vehicles in the queue.

Therefore

$$t = Nj/av$$

where v = free running speed of the traffic,

$\quad t$ = time for a vehicle to travel the length of the queue at the running speed.

The number of vehicles that arrive in this time is

$$\frac{qNj}{av}$$

and this correction should be added to the previously derived values of N giving

$$N = q\left(\frac{r}{2} + d\right)\left(1 + \frac{qj}{av}\right) \quad \text{or} \quad qr\left(1 + \frac{qj}{av}\right) \tag{45.3}$$

whichever is the greater.

Problems

The flows and saturation flows, given in Part A below, are observed on traffic-signal approaches. Select from Part B the appropriate value of the queue at the beginning of the green period for each of the approaches described in Part A. In all cases the inter-green period is 4 s, starting delays may be taken as 2 s of each green period and the spacing of queued vehicles may be taken as 6 m. Average speed on the approach is 6 m/s.

Part A

a) A two-lane approach with a width of 7·65 m carries a flow of 1000 p.c.u./h; the cycle time is 100 s and the actual green time 40 s. On this approach it may be assumed that vehicles arrive with a uniform headway distribution. An average vehicle is equal to ·2 p.c.u.

b) A single right-turning lane with a saturation flow of 1600 p.c.u./h carries a flow of 600 p.c.u./h; the actual green time is 60 s and the cycle time is 100 s. An average vehicle is equal to 1·3 p.c.u.

Part B

a) Approximately 8 vehicles.
b) Approximately 14 vehicles.
c) Approximately 20 vehicles.

Solutions

a) The saturation flow on this approach is given by

$$525 \times w \text{ p.c.u./h} = 525 \times 7·65 \quad \text{(see page 282)}$$
$$= 4015 \text{ p.c.u./h}$$

The capacity of the approach is given by

$$\frac{\text{effective green time} \times \text{saturation flow}}{\text{cycle time}} \text{ p.c.u./h}$$

and

effective green time = actual green time + amber period

$$- \text{ starting delays} \quad \text{(see chapter 37)}$$

The capacity of the approach

$$= \frac{41 \times 4015}{100}$$
$$= 1646 \text{ p.c.u./h}$$

The flow on the approach is 1000 p.c.u./h.

The flow on the approach is less than the capacity and because the vehicles arrive at regular intervals the maximum queue at the commencement of the green period is equal to the number of vehicles arriving during the preceding red and red-plus-amber periods.

The average queue at the commencement of the green period is then

$$N_u = qr \qquad \text{(see equation 45.1)}$$

$$= \frac{1000 \times (100 - 41)}{3600} \text{ p.c.u.}$$

$$= 16 \cdot 4 \text{ p.c.u.}$$

$$= 16 \cdot 4/1 \cdot 2 \text{ vehicles}$$

$$= 13 \cdot 6 \text{ vehicles}$$

(b) The capacity of the approach $= \dfrac{61 \times 1600}{100}$ p.c.u./h

$$= 976 \text{ p.c.u./h}$$

The flow on the approach $= 800$ p.c.u./h

The approach is nearing saturation so that the average number of vehicles waiting on the approach at the commencement of the green period is given by equation 45.3

$$N = q \left(\frac{r}{2} + d \right) \left(1 + \frac{qj}{av} \right) \qquad \text{or} \qquad qr \left(1 + \frac{qj}{av} \right)$$

whichever is the greater.

The average delay on the approach d can be calculated using the method described in section 43

$$d = cA + \frac{B}{q} - C \qquad \text{(equation 43.1)}$$

and

$$\lambda = 0 \cdot 61$$

$$x = 800/976 = 0 \cdot 82$$

$$M = \frac{800 \times 100}{3600 \times 1 \cdot 3} \text{ vehicles/cycle}$$

$$= 17 \cdot 1 \text{ vehicles/cycle}$$

By interpolation from table 43.1

$$A = 0 \cdot 153$$

By interpolation from table 43.2

$$B = 1 \cdot 87$$

By interpolation from table 43.3

$$C = 13 \text{ per cent}$$

$$d = 100 \times 0.153 + \frac{1.87}{0.17} - 13 \text{ per cent of previous two terms}$$

$$= 22.9 \text{ s}$$

Then

$$N = 0.17 \left(\frac{39}{2} + 22.9 \right) \left(1 + \frac{0.17 \times 6}{1 \times 6} \right)$$

$$\text{or} \qquad 0.17 \times 39 \left(1 + \frac{0.17 \times 6}{1 \times 6} \right) \qquad \text{(equation 45.3)}$$

$$= 6.2 \qquad \text{or} \qquad 7.8$$

that is

$$N = 7.8 \text{ vehicles}$$

46

The co-ordination of traffic signals

When several traffic signal-controlled intersections occur along a major traffic route, some form of co-ordination is necessary to prevent, so far as is possible, major road vehicles stopping at every intersection. Alternatively, or in addition, the signals may be co-ordinated to minimise delays to vehicles. Sometimes linking between signals is carried out to prevent queues stretching back from one intersection to the preceding signals.

Several forms of linking between signals are possible, three of which are the simultaneous system, the alternate system and the flexible progressive system. In these systems co-ordination between intersections is usually achieved by means of a master controller.

In the simultaneous system all signals along the co-ordinated length of highway display the same aspect to the same traffic stream at the same time. Some local control is possible using vehicle actuation but a master controller keeps all the local controllers in step and imposes a common cycle time. An obvious disadvantage of this control system is that drivers are presented with several signals each with a green aspect, and there is a tendency to travel at excessive speeds so as to pass as many signals as possible before they all change to red. Where turning traffic is light and intersections are closely spaced, then this system may have advantages for pedestrian movement.

The alternate system allows signal installations along a given length of road to show contrary indications. This means that if a vehicle travels the distance between intersections in half the cycle time, then a driver need not stop. The cycle time must be common to all the signals and must be related to the speed of progression. Major roads with unequal distances between intersections present difficulties for this reason.

With the flexible progressive system the green periods at adjacent intersections are offset relative to each other according to the desired speed on the highway. Progression along the highway in both directions must be considered and this usually results in a compromise between the flow in both directions and also between major and minor road flows.

A master controller keeps the local controllers, which may be either fixed-time or vehicle actuated, in step. Vehicle-actuated control is possible when there is no longer

ntinuous demand from all detectors, the signals changing in accordance with traffic
riving at the isolated intersection.

Any change of right of way must however not interfere with the progressive plan so
at at certain periods a change of right of way cannot take place because there would
insufficient time in which to regain right of way as required by the progressive plan.

Under very light traffic conditions, the flexible progressive system is likely to
oduce greater delays than an unlinked system because of the overriding priority of a
all number of major road vehicles. In some circumstances traffic density measuring
vices bring the master controller into action to impose an overall flexible progressive
tem as traffic volume increases, while at lower volumes the signals at each intersection
in an independent manner.

At key intersections vehicle actuation is often allowed and the control of the
ceding or following signals linked to the operation of the key intersection.

Where traffic leaving the key intersection is assisted to pass through the next inter-
tion then this is referred to as forward linking. If however traffic arriving at the key
ersection on a particular approach is given preference this is referred to as backward
king. Forward linking prevents a key intersection from becoming blocked while
kward linking will prevent a queue stretching back from the key intersection causing
o be blocked.

blem

najor traffic route through a central city area has frequent regularly spaced traffic
al intersections. It is proposed to co-ordinate these signals so as to minimise delay
ll vehicles passing through the intersections. Arrange the following control systems
rder of their suitability for the above situation.

A flexible progressive control system with fixed-time operation at each intersection.

A simultaneous-control system with a major road green time of 90 s and a cycle
e of 120 s.

An alternate-control system with vehicle actuation at each intersection.

A control system that imposes flexible progression during periods of heavier traffic
v and allows isolated-vehicle actuation at other times.

tion

order of merit of these control systems is:

This system would allow progression during periods of heavier flow but at other
s would allow the intersections to function under isolated vehicle actuated control
eal with fluctuations in the traffic flow.

This system would allow progression provided the cycle time was correctly chosen
at periods of low demand vehicle actuation would allow flexibility.

This system would offer progression at one level of flow but would be inflexible
er changing traffic conditions.

This system would not form a satisfactory method of control being unable to meet
ging traffic demands and being likely to produce long delays at low traffic flows.

47

Time and distance diagrams for linked traffic signals

When the flexible progressive system of co-ordinating traffic signals is employed, the it is frequently desirable to construct a time-and-distance diagram to estimate the bes offset or difference in the start of the green time, between adjacent signals.

The time and distance diagram is simply a graph on which time and hence the sign settings is plotted horizontally.

Distance along the major route between intersections is plotted vertically. The slo of any line plotted on this diagram represents the speed of progression along the maj route.

Before the diagram can be prepared it is necessary to examine each of the intersec which it is desired to co-ordinate and calculate the optimum cycle times for the expe traffic flows using the relationship

$$C_{\text{pract}} = \frac{0 \cdot 9L}{0 \cdot 9 - Y} \quad \text{or} \quad C_0 = \frac{1 \cdot 5L + 5}{1 - Y} \qquad \text{(see page 312 or 294)}$$

From these calculations it is possible to determine the intersection that is most heavily loaded and that therefore requires the longest cycle time. This intersection is then referred to as the key intersection and because in any linked system it is necess for each cycle time to be the same, or a multiple of the same value, the cycle time o the key intersection is adopted as the cycle time for the whole system.

As traffic flows vary throughout the day and also throughout the week, it may be necessary to employ differing key intersection and common cycle times and hence differing progression plans for differing days and times.

With the cycle time c_1 for the system determined, it is possible to calculate the effective green times for the differing phases at the key intersection from the relatio ship

$$\text{effective } g_1 = \frac{y_1}{Y}(c_1 - L) \qquad \text{(see page 299)}$$

The actual green time may then be determined if the lost time within the green period is known (see page 292).

344

The actual green time at the key intersection gives the minimum actual green time at the other intersections for the main route with progression.

To obtain the maximum green times for the intersections other than the key intersection it is necessary to determine the shortest acceptable green times for side road phases. The shortest acceptable effective green time for a side road phase is obtained from

$$\frac{y_{side} \times c_1}{0 \cdot 9}$$

and from a knowledge of the lost time in the green period the shortest acceptable actual green time can once again be calculated.

The longest actual green for the route with progression is then the linked cycle time c_1 minus the shortest acceptable actual green minus the intergreen periods.

With the calculated value of green times at the key intersection and the maximum and minimum green times at the other intersections known, it is possible to arrange the relative timing or offsets of the signals. In the preparation of this diagram the speed of progression between intersections should be chosen taking into account the known or likely speed/volume relationship for the highway and such physical features as horizontal and vertical curvature and pedestrian activity. Turning movements and critical queue lengths also should be considered in the arrangement of the progression through the intersection. A typical time/distance diagram is shown in figure 47.1.

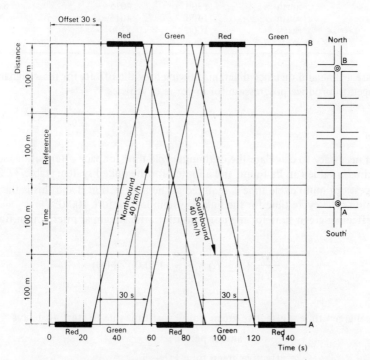

Figure 47.1 A time/distance diagram showing how a time offset is arranged between intersections A and B to give progression through two signals

Problem

Four two-phase traffic signal-controlled intersections along a major north/south traffic route are spaced at distances of 0·5 km apart. Details of the evening peak hour traffic flows at these intersections are given in table 47.1. Starting delays in all cases may be taken as 2 s of each green period.

TABLE 47.1

Intersection	Approach	Flow	Saturation flow	Lost time per cycle (s)
A	north	1250	4015	6
	south	1450	4015	
	east	1000	2250	
	west	800	1950	
B	north	1350	4015	8
	south	1550	4015	
	east	1200	2700	
	west	650	2250	
C	north	1100	4015	8
	south	1500	4015	
	east	900	2250	
	west	550	2250	
D	north	1300	4015	8
	south	1400	4015	
	east	1000	2700	
	west	600	1950	

Prepare a time and distance diagram showing how the offsets for these signals may be arranged to produce progression for major road vehicles.

Solution

It is first necessary to calculate the optimum cycle time for each intersection, so that the maximum value can be found and adopted for the linked system (table 47.2). The longest optimum cycle time is required at the intersection B and this is the key intersection. The cycle time for the linked system is therefore 100·0 s.

The effective green time for the north/south traffic stream at intersection B is given by

$$\frac{y_N/S}{Y}(c_1 - L) = \frac{0·39}{0·83}(100·0 - 8·00)$$

$$= 43·2 \text{ s}$$

and actual green time = effective green time + lost time due to starting delays

$$- \text{ amber period}$$

$$= \text{effective green time} + 2 - 3$$

$$= 42·2 \text{ s}$$

TABLE 47.2

Intersection	Approach	y value	C_0
A	north	0·31	
	south	0·36	
	east	0·44	$\dfrac{1\cdot5 \times 6 + 5}{1 - (0\cdot36 + 0\cdot44)} = 70\cdot0 \text{ s}$
	west	0·41	
B	north	0·34	
	south	0·39	
	east	0·44	$\dfrac{1\cdot5 \times 8 + 5}{1 - (0\cdot39 + 0\cdot44)} = 100\cdot0 \text{ s}$
	west	0·29	
C	north	0·27	
	south	0·37	
	east	0·40	$\dfrac{1\cdot5 \times 8 + 5}{1 - (0\cdot37 + 0\cdot40)} = 73\cdot9 \text{ s}$
	west	0·25	
D	north	0·32	
	south	0·35	
	east	0·37	$\dfrac{1\cdot5 \times 8 + 5}{1 - (0\cdot35 + 0\cdot37)} = 60\cdot7 \text{ s}$
	west	0·31	

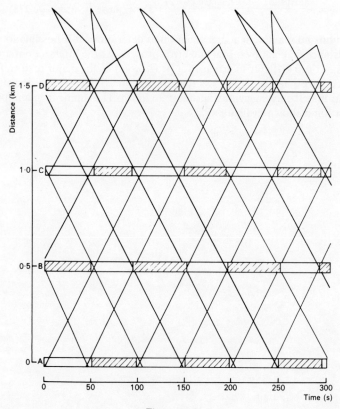

Figure 47.2

This is also the minimum actual green time for the north/south traffic stream at the remaining intersections.

The maximum actual green times at the remaining intersections are then calculated from a consideration of the minimum effective green time required for side road traffic using the relationship

$$\frac{y_{side} \times c_1}{0.9}$$

(1)	(2)	(3) Minimum effective green on minor route $\dfrac{Y_{side} \times c_1}{0.9}$	(4) Minimum actual green on minor route (3) − 1	(5) Maximum actual green on major route $c_1 - (4) -$ intergreen time*
Intersection	y_{side} max	(s)	(s)	(s)
A	0.44	48.9	47.9	44.1
C	0.40	44.4	43.4	46.6
D	0.37	41.1	40.1	49.9

* Intergreen time = $\dfrac{(\text{lost time} - \text{starting delays} \times 2)}{2}$ + 3 s per phase   (see page 278).

With minimum and maximum actual green times determined it is possible to plot a time and distance diagram by a trial and error process. As the intersections are regularly spaced, it is not too difficult to arrange for progression of the major road traffic stream in both directions. The diagram allows for progression, so that vehicles travel between intersections in half a cycle giving a speed of 36 km/h. The timing diagram for this very simple situation is shown in figure 47.2.

48

Platoon dispersion and the linking of traffic signals

trial and error approach such as is involved in the preparation of a time-and-distance diagram can produce reasonable progression along a major traffic route. Where however it is desired to minimise delay at signal-controlled intersections in a network a more rigorous approach is desirable.

An alternative method of linking traffic signals so as to minimise delay is to consider the one-way stream of vehicles which after release from one traffic signal travel to the next controlled intersection and are then released when the signals change to green.

If the cumulative demand function, or the number of vehicles arriving at a traffic-signal approach, and the cumulative service function, or the number of vehicles dis-

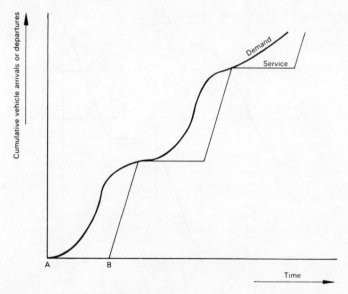

Figure 48.1 Cumulative demand and service volumes on a traffic signal approach

charging from a traffic-signal approach, are known then the area between the curves gives the delay on the approach. These functions are illustrated in figure 48.1.

The vertical distance between the curves represents the number of vehicles delayed at the stop line at any instant while the horizontal distance between the curves represents the duration of delay.

If the time interval between the arrival of the first vehicle at the stop line (A) and t departure of the first vehicle from the stop line (B) is varied then the area between the curves and hence the total delay is varied. By adjustment of the time interval A–B the delay may be minimised.

Such an approach can be used to minimise delay at intersections provided the form of the demand and service functions are known. There is little difficulty in defining the service function for it is simply illustrated by variations in the flow across the stop line (see page 291).

It is the form of the demand function that must be defined and in a linked-signal system the problem is to predict how platoons of vehicles that are released from a traffic-signal approach disperse as the vehicles travel down the highway towards the next stop line.

Problem

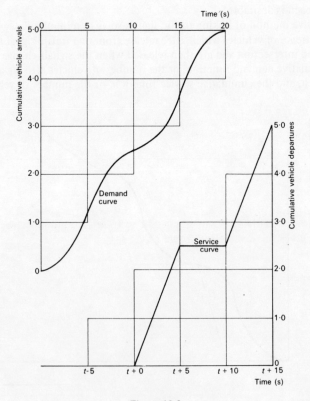

Figure 48.2

The cumulative demand function and the cumulative service functions for a traffic-signal approach are given in figure 48.2. Is the optimum time t for the commencement of green so as to minimise delay 0, 1·5, 2·5 or 4·5 s?

Solution

The offset of the green signal in relation to the arrival of the first vehicle at the stop line so as to minimise delay can be obtained by the graphical superposition of the service curve on the demand curve. In doing this it is necessary to minimise the area between the curves.

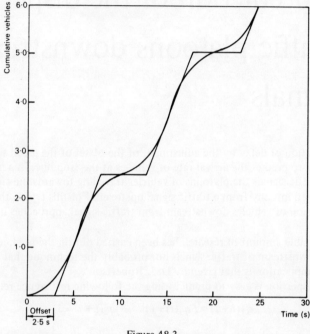

Figure 48.3

This is shown in figure 48.3 where it can be seen that for minimisation of the area between the curves the time offset should be 2·5 s.

49

The prediction of the dispersion of traffic platoons downstream of signals

The minimisation of delay by the adjustment of the offset of the green signal depends on the ability to predict the arrival rate of vehicles at the stop line. In a network controlled by traffic signals the platoons of vehicles travelling towards the stop line will have been discharged initially from a traffic signal upstream. For this reason the prediction of the dispersion of vehicles downstream from traffic-signal approaches is of considera importance.

A considerable amount of research has been carried out on the prediction of platoon dispersion downstream of traffic signals but probably the technique that has received the greatest application is that given by D. J. Robertson[1].

Platoon dispersion is easy to predict using the following recurrence relationship

$$q_2(i + t) = Fq_1(i) + (1 - F)q_2(i + t - 1) \tag{49.1}$$

where $q_2(i)$ is the derived flow in the ith time interval of the predicted platoon at a point along the road,
$q_1(i)$ is the flow in the ith time interval of the initial platoon at the stop line; t 0·8 times the average journey time over the distance for which the platoon dispersion is being calculated (measured in the same time intervals used for $q_1(i)$ a $q_2(i)$, which is usually 1/50th of the cycle time),
F is a smoothing factor.
Using data from four highway sites in London an expression for F has been given a

$$F = 1/(1 + 0·5t) \tag{49.2}$$

Reference

1. D. J. Robertson. Transyt: a traffic network study tool. *Road Research Laboratory Report* 253 (1969)

Problems

A highway link commences and terminates with traffic signal-controlled intersections. The discharge from the first intersection may be assumed to have a uniform rate of 1 vehicle/unit time and commences at zero time. The green time at the first intersection is 5 time units and at the second intersection 8 time units. The average travel time of the platoon between intersections may be taken as 10 time units and the travel time of the queue leader as 8 time units.

Using the recurrence relationship derived by Robertson is the percentage of vehicles delayed at the end of the green period, when the offset of the second signal from the first signal is 8 time units, approximately 25 per cent, 27·5 per cent, 35 per cent?

Solutions

The average journey time is given as 10 time units. Then

$$t = 0·8 \times \text{average journey time}$$
$$= 8 \text{ time units}$$

and

$$F = 1/(1 + 0·5t) \quad \text{(see equation 49.2)}$$
$$= 0·2$$

Using the recurrence relationship given in equation 49.1

$$q_2(i + t) = Fq_1(i) + (1 - F)q_2(i + t - 1)$$

TABLE 49.1

$q_1(i)$	$i + t$	$q_2(i + t)$	Cumulative Demand	Cumulative Service
1·0	9	$0·2 \times 1·0 + 0·8 \times 0 \quad = 0·20$	0·20	1·0
1·0	10	$0·2 \times 1·0 + 0·8 \times 0·2 = 0·36$	0·56	2·0
1·0	11	$0·2 \times 1·0 + 0·8 \times 0·36 = 0·49$	1·05	3·0
1·0	12	$0·2 \times 1·0 + 0·8 \times 0·49 = 0·59$	1·64	4·0
1·0	13	$0·2 \times 1·0 + 0·8 \times 0·59 = 0·67$	2·31	5·0
	14	$0·2 \times 0 + 0·8 \times 0·67 \quad = 0·54$	2·85	6·0
	15	$0·2 \times 0 + 0·8 \times 0·54 \quad = 0·43$	3·28	7·0
	16	$0·2 \times 0 + 0·8 \times 0·43 \quad = 0·34$	3·62	8·0

Total platoon content = initial discharge x time

$$= 1·0 \text{ vehicles/time unit} \times 5 \text{ time units}$$
$$= 5·0 \text{ vehicles.}$$

Cumulative demand during green period at the second intersection is 3·62 vehicles.

Number of vehicles delayed = 5·0 − 3·62 = 1·38 vehicles

$$= 27·5 \text{ per cent (approximately)}$$

50

The delay/offset relationship and the linking of signals

By the use of the technique of calculating the delay to vehicles on a traffic-signal approach from the difference between the demand and service distributions, as discuss in chapter 48, and the prediction of the demand distribution from a knowledge of platoon diffusion relationships as described in chapter 49, it is possible to obtain a relationship between the offset of one signal relative to another and the delay to vehicles passing through both signals.

Using this approach it is possible to calculate the offset or difference in time betwee the commencement of green for two linked signals likely to result in minimum delay. The process can be extended using a method proposed by the Road Research Laborat to a highway network. In this way a series of signal offsets can be obtained for differe sets of traffic flow conditions. If signal indications are changed by a central controller or computer then when changes in traffic flow take place it will be possible to select t appropriate offset sequence to minimise delays to vehicles travelling through the netw

A method of using delay/offset relationships to obtain the relative timings or offse of traffic signals in a network has been proposed by P. D. Whiting of the Transport an Road Research Laboratory and is referred to as the Combination method.

The basic assumptions made in the Combination method are:

(i) The settings of the signals do not affect the amount of traffic or the route used.
(ii) All signals have a common cycle (or have a cycle that is a submultiple of some master cycle).
(iii) At each signal the distribution of the available green time among the phases is known.
(iv) The delay to traffic in one direction along any link of the network depends solely on the difference between the settings of the signals at each end of the link; it is not affected by any other signals in the network.

(The term 'link' is used to refer to the length of road between signalised intersectic because this is the more familiar usage when discussing networks. Thus 'link' and 'section' are broadly synonymous.) It is assumed unless otherwise stated that the follc ing basic data relating to the network is available:

(i)   the cycle time to be used for the network;

354

(ii) the delay/difference-of-offset relation in each direction for each link. The delay/difference-of-offset relation comprises N delays, one for each of the N possible differences of offset between the signals at each end of the link.

Delays in links can be combined in two ways:

. In parallel.
. In series.

Consider two one-way links in parallel between signalised intersections as illustrated in figure 50.1 for which the delay/offset relations are also given.

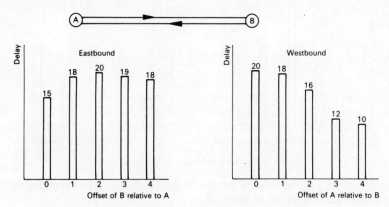

Figure 50.1 Delay/offset relationships for link AB

The combined delay/difference of offset relationship for the link can then be obtained by addition, taking care to add delays so that they are relative to the same signal. The total relationship is shown in figure 50.2.

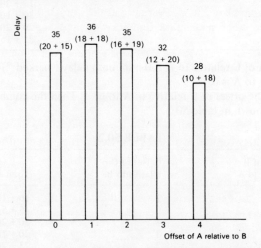

Figure 50.2 The combined delay/offset relationship for link AB

It can be seen from this combined relationship that minimum delay occurs when th
offset of A relative to B is 4.

When links are arranged in series, the delay difference in offset relationship for each
difference of offset between the extremities has to be calculated and the minimum val
selected. Consider the two links in series AB and BC shown in figure 50.3 for which th
delay/offset relationships are also given.

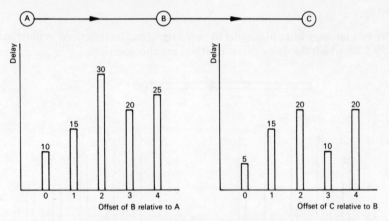

Figure 50.3 The combination of links in series

Consider initially the offset of C relative to A to be 0. Then the combined delay/
offset relationship is as shown in table 50.1.

TABLE 50.1

Offset of B relative to A	Offset of C relative to B	Total delay
0	0	10 + 5 = 15*
1	4	15 + 20 = 35
2	3	30 + 10 = 40
3	2	20 + 20 = 40
4	1	25 + 15 = 40

When the offset of C relative to A is 0 minimum delay (marked *) occurs when the
offset of B relative to A is 0.

Consider next the offset of C relative to A to be 1. Then the combined delay/offset
relationship is as shown in table 50.2.

TABLE 50.2

Offset of B relative to A	Offset of C relative to B	Total delay
0	1	10 + 15 = 25
1	0	15 + 5 = 20*
2	4	30 + 20 = 50
3	3	20 + 10 = 30
4	2	25 + 20 = 45

When the offset of C relative to A is 1 minimum delay (marked *) occurs when the offset of B relative to A is 1.

Consider next the offset of C relative to A to be 2. Then the combined delay/offset relationship is as shown in table 50.3.

TABLE 50.3

Offset of B relative to A	Offset of C relative to B	Total delay
0	2	10 + 20 = 30*
1	1	15 + 15 = 30*
2	0	30 + 5 = 35
3	4	20 + 20 = 40
4	3	25 + 10 = 35

When the offset of C relative to A is 2 minimum delay (marked *) occurs when the offset of B relative to A is 2.

Consider next the offset of C relative to A to be 3. Then the combined delay/offset relationship is as shown in table 50.4.

TABLE 50.4

Offset of B relative to A	Offset of C relative to B	Total delay
0	3	10 + 10 = 20*
1	2	15 + 20 = 35
2	1	30 + 15 = 45
3	0	20 + 5 = 25
4	4	25 + 20 = 45

When the offset of C relative to A is 3 minimum delay (marked *) occurs when the offset of B relative to A is 0.

Finally when the offset of C relative to A is 4, the combined delay/offset relationship is as shown in table 50.5.

TABLE 50.5

Offset of B relative to A	Offset of C relative to B	Total delay
0	4	10 + 20 = 30
1	3	15 + 10 = 25*
2	2	30 + 20 = 50
3	1	20 + 15 = 35
4	0	25 + 5 = 30

When the offset of C relative to A is 4 minimum delay (marked *) occurs when the offset of B relative to A is 1.

It is now possible to produce a combined delay/difference of offset relationship, as given in figure 50.4.

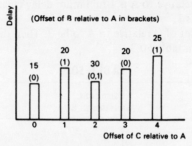

Figure 50.4 The combined delay/offset relationships for links AB and BC

It can be seen that minimum delay for the combined link occurs when the offset of C relative to A is 0 and when the offset of B relative to A is 0 respectively.

Problem

For the network and delay/offset relationships shown in figure 50.5 determine the offsets for minimum delay in the network.

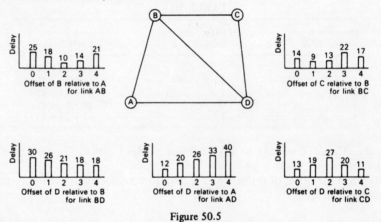

Figure 50.5

Solution

Combination of links BC and CD. (A series combination.)

1. *Difference of offset between D and B of 0*

Offset of C relative to B	Offset of D relative to C	Total delay
0	0	14 + 13 = 27
1	4	9 + 11 = 20*
2	3	13 + 20 = 33
3	2	22 + 27 = 49
4	1	17 + 19 = 36

2. *Difference of offset between D and B of 1*

Offset of C relative to B	Offset of D relative to C	Total delay
0	1	14 + 19 = 33
1	0	9 + 13 = 22*
2	4	13 + 11 = 24
3	3	22 + 20 = 42
4	2	17 + 27 = 44

3. *Difference of offset between D and B of 2*

Offset of C relative to B	Offset of D relative to C	Total delay
0	2	14 + 27 = 41
1	1	9 + 19 = 28
2	0	13 + 13 = 26*
3	4	22 + 11 = 33
4	3	17 + 20 = 37

4. *Difference of offset between D and B of 3*

Offset of C relative to B	Offset of D relative to C	Total delay
0	3	14 + 20 = 34
1	2	9 + 27 = 36
2	1	13 + 19 = 32
3	0	22 + 13 = 35
4	4	17 + 11 = 28*

5. *Difference of offset between D and B of 4*

Offset of C relative to B	Offset of D relative to C	Total delay
0	4	14 + 11 = 25*
1	3	9 + 20 = 29
2	2	13 + 27 = 40
3	1	22 + 19 = 41
4	0	17 + 13 = 30

The minimum delay values are marked * and for the varying differences of offset between D and B the delay histogram can be plotted when BC and CD are combined (figure 50.6).

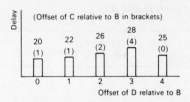

Figure 50.6 The series combination of links BC and CD

Combination of above and link BD (a parallel combination). The delay for link BD is known for the offset of D relative to B and so the combination of the two links is a straightforward addition, giving figure 50.7.

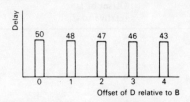

Figure 50.7 The parallel combination of link BD and the previous combination

Combination of above and AB (a series combination)

6. *Difference of offset between D and A of 0*

Offset of D relative to B	Offset of B relative to A	Total delay
0	0	50 + 25 = 75
1	4	48 + 21 = 69
2	3	47 + 14 = 61
3	2	46 + 10 = 56*
4	1	43 + 18 = 61

7. *Difference of offset between D and A of 1*

Offset of D relative to B	Offset of B relative to A	Total delay
0	1	50 + 18 = 68
1	0	48 + 25 = 73
2	4	47 + 21 = 68
3	3	46 + 14 = 60
4	2	43 + 10 = 53*

8. *Difference of offset between D and A of 2*

Offset of D relative to B	Offset of B relative to A	Total delay
0	2	50 + 10 = 60
1	1	48 + 18 = 66
2	0	47 + 25 = 72
3	4	46 + 21 = 67
4	3	43 + 14 = 57*

9. *Difference of offset between D and A of 3*

Offset of D relative to B	Offset of B relative to A	Total delay
0	3	50 + 14 = 64
1	2	48 + 10 = 58*
2	1	47 + 18 = 65
3	0	46 + 25 = 71
4	4	43 + 21 = 64

10. *Difference of offset between D and A of 4*

Offset of D relative to B	Offset of B relative to A	Total delay
0	4	50 + 21 = 71
1	3	48 + 14 = 62
2	2	47 + 10 = 57*
3	1	46 + 18 = 64
4	0	43 + 25 = 68

The minimum values are marked *. For varying offsets between D and A the delay histogram can be plotted (figure 50.8) when AB is added to the previous combinations.

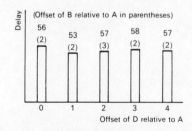

Figure 50.8 The series combination of link AB and the previous combination

Finally the previous combinations can be combined with the link AD (a parallel combination).

11. The combination is a straightforward addition giving the histogram of figure 50.9.

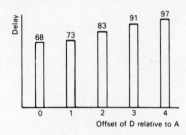

Figure 50.9 The parallel combination of link AD and the previous combination

Minimum delay for the total combination thus occurs when the offset of D relative to A is 0 (para 11).

When the offset of D relative to A is 0 minimum delay occurs when the offset of D relative to B is 3 and the offset of B relative to A is 2 (para 6).

When the difference of offset between D and B is 3 minimum delay occurs when offset of C relative to B is 4 and the offset of D relative to C is 4 (para 4).

51

Some area traffic control systems

During the late 1950s proposals were made that the linking and co-ordination of traffic signals which already existed in a linear manner along many major highways could be extended to whole areas.

One of the earliest schemes in which a digital computer was used to control a number of traffic signal-controlled intersections was a pilot project begun in Toronto in 1959.

Before that time all the traffic control signals operated on a pretimed basis with only a small amount of co-ordination. At this time the Traffic Research Corporation suggested that the use of a digital computer operated in real time and on line could

. Receive traffic information from a large number of vehicle detectors, determine the time intervals required at each individual intersection, and optimise these for overall system efficiency taking account of existing conditions.
. Determine the optimum time offsets between individual intersections, considering the existing traffic speed and direction of flow.
. Control the individual signals directly to produce optimum conditions.
. Check the signal operation and resulting traffic movement to ensure that optimum conditions have been obtained.

This report pointed out that a large digital computer could carry out all these functions with sufficient speed to ensure that the control system could be varied almost instantaneously. In addition any type of existing traffic control system could be duplicated with simple programming and the system, timing and offsets could be varied from the control centre at will.

At this early stage in the development of area control, some doubts were expressed about the feasibility of the scheme. A smaller initial pilot project was initiated therefore in 1959 in which some fifteen signals were controlled in various ways between pretimed systems and fully traffic responsive operation.

So successful was this project that it was decided to extend computer control to the full system. Delivery and installation of equipment began in 1963 and in 1967 some 100 signals on major arterial roads were under effective traffic responsive control. 250 further signals were incorporated during 1967 and new installations were thereafter expected to be connected at a rate of between 30 and 40 per year.

In this control system the basic function of the computers is to inspect each individual signal approximately once per second and to determine if a change of aspect is required by a comparison between the elapsed time of the current aspect and the time

363

considered necessary to satisfy the needs of the alternating traffic flows at the inter-section and the system as a whole.

Under some control systems required green times were predetermined for efficient operation of the system as a whole but it was considered that this made for inflexible operation. To overcome this disadvantage, traffic-responsive systems calculate the optimum duration of each interval in one of a number of different ways, which depend upon the physical characteristics of the junction and actual traffic conditions. As each junction is likely to require different treatment, the operation of the computer is controlled by a plan that specifies for each signal the exact factors to be considered.

In the British Isles little published information was available in the early 1960s on which the benefits of area traffic control systems could be judged and for this reason it was decided to set up two experimental area traffic control schemes in Glasgow and in West London.

West London System

Two experimental area traffic control schemes set up in the British Isles which have been extensively reported, are the Glasgow and West London schemes.

The area selected in West London covers an area of some 6 square miles bounded by the Thames on the south, Hyde Park on the north, to the east by Sloane Street and on the west by Fulham Palace Road. Within this area all the signals were already vehicle actuated and some were already linked. In particular a linked system already operated along Cromwell Road, which carried 38 000 vehicles a day towards central London.

Because it was necessary to provide a traffic control system that was at least equal to that which already existed, two-level control was used. The existing intersection controllers were retained and these are used to control the intersections subject to the overriding control of the central computer.

Control has been referred to as strategic control and tactical control[1]. Strategic or overall control determines:

1. Which intersections should be allowed to operate independently of the computer, control being exercised by the intersection controller.
2. Which sub-areas of the total control area should be progressive.
3. Which progression plans should be selected for each sub-area.
4. Which sub-areas should be amalgamated to form larger sub-areas.
5. Which progression plans should be selected for each of these larger sub-areas.
6. Whether modification of the progression plans is required under heavy traffic conditions.

Strategic control is the repeated answering of these questions, decisions being made on the basis of the overall traffic pattern and not on the needs of individual vehicles. When the central controller has made a decision by analysing the traffic pattern this decision will be valid until the computer detects a change in overall traffic flows.

Tactical control, on the other hand, deals with the short-term aspects of traffic flow and even with the demands of individual vehicles. Decisions by the computer however exclude those made by the local controllers. It is planned that tactical decisions will include:

1. Phase splitting (that is, limitation of permitted movement to one direction only).
2. Retaining a green signal in congested conditions.
3. Linking of adjacent controllers when required.
4. Varying the commencement time of a phase as scheduled by the progressive plan in use.
5. Adjusting cycle time.
6. Introducing and removing additional phases.
7. Banning and permitting certain phases.
8. Introducing short cycles to clear intersections.
9. Introducing and retaining of certain phases for emergency purposes.
10. Introducing and removal of diversion routes.
11. Introducing and removing tidal-flow schemes.

Mitchell[1] has stated that the main intention is to establish progression over the greatest route mileage practicable whenever traffic conditions are favourable. The progression plans will be selected approximately every ten minutes from the plan library. On the selected plan tactical short-term measures will be superimposed to give flexibility to the control arrangements.

Glasgow System

Parallel with the West London experimental scheme the Road Research Laboratory has been carrying out experimental work on the control of 80 traffic signals dealing with traffic over a square mile in the centre of Glasgow. Their object was to measure the benefit of various systems of co-ordinated control.

Hillier[2] has stated that the following traffic control systems would be assessed in the experiment:

1. Fixed-time progressions based on historical knowledge of traffic conditions selected from a library by time of day, with some provision for manual selection according to local conditions.
2. The 'PR' system in which a similar library of fixed-time programs is available, but the selection is based on an assessment of traffic conditions at representative points in Glasgow.
3. Isolated operation by vehicle actuation.
4. The 'commencement of maximum' system, which was used to link about 45 of the signals on main roads in central Glasgow.
5. The flexible progressive systems for linking vehicle-actuated signals.
6. A saturation flow mode of control applied independently to each intersection.
7. A form of signal control that maintains an equal degree of saturation on each phase of an individual intersection.
8. A form of control based on knowledge or prediction of traffic flows expected on each approach up to one cycle ahead and designed to select the time of the next phase change so as to minimise total delay at the intersection.

Control systems (1) and (2) use libraries of progressive fixed time settings for minimum journey time as previously described. Control system 3 is the normal vehicle-actuated system used in the United Kingdom.

The commencement of maximum system (4) was already in use in Glasgow to control about 45 signals on the main streets in the central area. It had a fixed-cycle master controller which supervised two phases of a local controller at each signal-controlled intersection. The operation of the local controller differs from the normal vehicle-actuated controller in the following manner:

1. In the commencement of maximum system the impulses given by a vehicle as it crosses a detector on a junction approach are sent both to the controller at that junction and to the controller at the next junction ahead in the line of travel.
2. In addition there is no normal maximum timing but instead the local controller receives two impulses, one for each of the two phases being supervised, from the master controller at predetermined times. The reception of a pulse terminates green on the appropriate phase, provided that the minimum green has elapsed and that the controller has received a demand for another phase. A pulse has no effect if the appropriate phase is not running when it is received.

In this system the signals change according to the local traffic demand when flows are light but in accordance with the master pulses when flows are heavy.

Control system (5) has master controllers that supervise the action of local controllers and exert overriding control at certain points in a common cycle to ensure progression of major road traffic. Under heavy traffic this system gives normal fixed time progression.

The saturation-flow mode (6) method of control is a form of isolated vehicle-actuated operation. A phase runs, subject to maximum and minimum greens, for as long as at least one of the roads controlled by that phase has traffic entering the intersection at the saturation flow rate. A statistical examination of the gaps in the flow indicates the termination of saturation flow conditions and allows a phase change to be made if desired.

Control mode (7) aims to maintain an equal degree of saturation on each phase. It operates in conjunction with fixed-time progression, allocating to the major and minor roads the green periods within the fixed cycle length. In one-way progression the offset of the centre of the major road green is fixed, any change in length being distributed equally between the beginning and end of the green period.

The final control mode (8) is based on predicted vehicle arrivals at major intersections that are likely to become critically loaded. The main object of the control scheme is to minimise delay by increasing capacity at these major intersections.

Using predicted arrivals at the stop line, the computer calculates whether it would be better to terminate the green signal at the end of the minimum green period or for the signal to remain at green for another second. No change of signal aspect takes place if the total delay over the whole cycle would be reduced by waiting. If the first calculation shows delay would be increased by waiting, then the calculation is repeated in 1 second steps until, if the delay is increased by waiting for 5 seconds, the signals change their aspect.

Computer control of the Glasgow central area signals began in the autumn of 1967 and the results of the use of four control schemes have been reported[3].

The first trial compared fixed-time progression schemes calculated by the Combination method with the local control that existed in Glasgow in the autumn of 1967. The combination progression schemes consisted of three plans for the morning peak, the off-peak and evening peak conditions prepared from historical data. The cycle times were chosen to give 10 per cent reserve capacity at the busiest junctions, subject to a maximum of 120 seconds. Green times were chosen to give an equal degree of saturation on each stage at a junction. The offset in time between the appearance of greens at adjacent intersections was calculated by the Combination method. It was concluded that the Combination settings produced a statistically significant improvement of 12 per cent in the average journey during the period covered by the survey.

A second trial was carried out in the spring of 1968 between the Combination method and Transyt. As with the Combination method three separate plans for each of the three periods of the day were used. It was noted that a further improvement of 4 per cent in average journey times over the Combination method was obtained by the Transyt method[3].

During the autumn of 1968 a third trial was made on Flexiprog, a vehicle-actuated flexible progressive system. The timings were selected using the Combination method to give the basic linking. The use of this system did not show any improvement over the Combination method fixed-time scheme[3].

In the spring of 1969 the equal degree of saturation system, Equisat was used to control the signals. The scheme used had the same cycle times and linking as the standard scheme but the allocation of green times could be varied automatically to equalise the degree of saturation on each stage[3].

The normal pneumatic detectors were used to measure both the flow and the saturation flow on each arm, preset figures being provided for use when the measured figures were rejected as unsuitable. Measured flows were obtained by taking the flow during each cycle and applying an experimental smoothing process to successive results.

A tentative measure of saturation flow was obtained by taking the maximum eight-second count of axles passing over the detector during the period from 6 to 22 seconds after the start of green. Unsuitable values were rejected by statistical tests. The figure used was a moving average of the last eight successive values of saturation flow which may either be acceptable measured values or the preset value substituted for a rejected measurement.

It was found that as with the Flexiprog the Equisat method did not produce any improvement on the Combination method. Holroyd and Hillier[4] have discussed these results. They considered the results surprising in that a method that allocated green time according to traffic conditions at the local controller should produce less delay than the fixed-time system on which it was based. They stated that this might still be true for Flexiprog in very low flow conditions but in the case of Equisat lightly trafficked intersections are unlikely to benefit from a redistribution of green times.

An additional mode of control referred to as Plident (platoon identification scheme) has also been tested in Glasgow. This scheme identifies the movement of platoons of traffic across the road network and attempts to operate the signals in a way that avoids delay to platoons of vehicles on priority routes. In this scheme only one approach to each signal may be a priority route, programmed to handle the greatest volume of traffic possible. In the Glasgow study area, where there are a number of one-way streets, about 40 per cent of the traffic travelled on priority routes[4].

The Plident control scheme is unique in that it does not have any fixed cycles or offsets, but adjusts the length of each stage to suit the traffic platoons. This is achieved by identifying platoons by the expected time of arrival of the front of the platoon at the stop line and also by noting the expected number of vehicles in the platoon.

In this system each stage normally continues until the end of saturation flow has been detected on all arms controlled by the stage, subject to a 180 s maximum. The end of saturation is detected by axle counts made as vehicles cross the detector on the approach. The level of saturation flow is determined from the estimated saturation flow the time since the start of green and the proportion of the platoon that is estimated to have passed the detector.

After the end of saturation has been detected on all arms of a stage, a change to the next stage begins immediately provided the minimum green period has expired and the next stage does not show green to the priority route. Should the next green show green to the major route then the change is delayed by an amount sufficient to ensure that the next platoon arrives at the stop line immediately after the lights have changed and any waiting vehicles have cleared. This avoids delay to the incoming platoon and also allows waiting vehicles to be added to the front so as to form a single platoon at the next junction downstream.

Additionally the stage preceding the priority stage may be terminated without reaching the 'end of platoon' stage, should this be necessary to avoid delaying the priority platoon, provided that the non-priority stage is estimated to require less than 15 s further green and that it ran to 'end of saturation' on the previous occasion.

When a non-priority platoon cannot be cut off then the priority platoon is given its green as soon as possible up to a maximum of 10 s late. After this time the priority platoon loses its priority and is added to the waiting vehicles to be held until the next priority platoon approaches. Should an 'end of platoon' occur on the stage preceding the priority stage, an immediate change to the priority stage is made without waiting for the next platoon, providing the controlling data shows that there is sufficient time to clear the waiting vehicles on the priority stage and service the other stages before a green is required for the next priority platoon.

Holroyd and Hillier stated that the effect of Plident is to give the priority routes just the amount of green they require at times which avoid delay to the approaching platoons. The non-priority streets are compensated to some extent for the lack of co-ordination of their green times with their platoon arrivals by being given the spare green time.

The Plident control scheme was tried out during the autumn of 1969 but it was noted that the delay was considerably greater than when the fixed time scheme was in use.

Current practice in area traffic control

The control systems which have been tested in urban areas may be grouped into three general types. Firstly control plans which are based on past records of traffic flows where the optimising process is carried out off-line, usually by the computer which issues the signal aspect change instructions. The optimising techniques which are comm employed are the Sigop, Combination and Transyt methods. The first of these is freque used in the United States while the last has become almost universally employed in Gre Britain.

The second general type of control is where signals respond individually to traffic detected on the approaches but subject to some form of underlying basic co-ordination Typical of these systems are Flexiprog and the Equisat co-ordination.

The third type comprises fully responsive systems in which signal settings are calcula on-line using traffic information from detectors on the highway. The Plident system is a scheme of this type.

As has been previously stated the most common form of control in Great Britain is obtained using a scheme of the first type where flows are optimised using the Transyt method. The latest version of this method is Transyt 6 which optimises the fixed time settings of a network of signals and allows bus travel time including time spent at bus stops to be taken into account in the optimising process.

The use of this optimising method has been described in considerable detail by Robertson and Gower[5] and by Holroyd.[6]

The first step is the selection of the network which is to be optimised: this will usually consist of all road sections between the traffic signals concerned. A separate link is assigned to each direction of travel, a two way road becoming two links.

Secondly a decision has to be made on the number of plans which will be required for the changing traffic flow conditions. As a good optimised plan is not considered to be very sensitive to reasonably large changes in traffic flow, Holroyd has suggested that normally a morning peak plan, an evening peak flow plan and an off-peak plan will be necessary with perhaps a lunch time peak plan. If control is to extend into the evenings, then an additional plan will be required together with plans for special events which produce unusual traffic flows. In total it is estimated that the maximum number of plans will not normally exceed ten. Frequent changes of plan should be avoided because a plan change is believed to add half a minute to the delay experienced by each vehicle.

As the traffic plan is based on historical data it is necessary to obtain details of journey times and speeds for the network. Great accuracy is not necessary and frequently estimates of the desired speed, unimpeded by queues, can be made. Actual vehicle flows at the intersections must be known as well as the saturation flows for each approach.

In the Transyt technique it is necessary to select a common cycle time for each area. It is recommended that for each intersection the practical cycle time (p. 344) be calculated and from a consideration of the values obtained it is necessary to select one of the following: a single cycle time for the whole area; double and single cycle times for the whole area; different cycle times for sub-areas and different double and single cycle times for sub-areas. Double cycling has advantages over a single long cycle when both short and long practical cycle times are calculated along a busy road with closely spaced intersections.

Once a cycle time or a double cycle time has been selected it is necessary to determine the distribution of green times at each intersection. This may be done manually to obtain an equal degree of saturation on opposing stages or the Transyt program itself can determine green time splits by a hill climbing process.

The latest Transyt program allows up to five different classes of vehicle to be considered. For each class of vehicle the average arrival flow profile is calculated using vehicle departure profiles. Vehicles are discharged over the stopline according to their arrival time which means that the saturation flow is shared. Stops and delays to vehicles are weighted in the program according to the estimated number of passengers.

References

G. Mitchell. Control concepts of the west London experiment. Symposium on Area Control Road Traffic, Institution of Civil Engineers, London (1967)

2. J. A. Hillier. Glasgow experiment in area traffic control. *Traff. Engng Control*, 7 (1965) 502–9 and 7 (1966), 569–71
3. J. Holroyd and J. A. Hillier. Area traffic control in Glasgow, a summary of results from four control schemes. *Traff. Engng Control*, 11 (1969) 220–3
4. J. Holroyd and J. A. Hillier. The Glasgow Experiment: Plident and after. Road Research Laboratory Report L.R. 384 (1971)
5. D. I. Robertson and P. Gower. User guide to Transyt version 6. Road Research Laboratory Report S.R. 255 (1977)
6. Joyce Holroyd. The Practical implementation of combination method and transyt programs. Road Research Laboratory Report L.R. 518 (1972)

APPENDIX

Definition of symbols used in Part 3

a = number of traffic lanes on a signal approach

A = $\dfrac{(1 - \lambda)^2}{2(1 - \lambda x)}$ = first term in the equation for average delay on a traffic-signal approach; due to the uniform rate of vehicle arrivals

B = $\dfrac{x^2}{2(1 - x)}$ = second term in the equation for average delay on a traffic-signal approach; due to the random nature of vehicle arrivals

B_r = mean minimum time headway between right-turning vehicles as they pass through a traffic signal-controlled intersection

B_0 = mean minimum time headway between opposing vehicles as they pass through a traffic signal-controlled intersection

B_1 = mean minimum time headway between bunched vehicles on the major road

B_2 = mean minimum time headway between vehicles as they discharge without conflict from the minor to the major road

c_l = common cycle time of linked signals

c = correction term in the equation for average delay on a traffic-signal approach

c_0 = optimum cycle time

c_m = minimum cycle time

d = average delay per vehicle on a traffic-signal approach

F = smoothing factor used in the prediction of platoon dispersion

g = effective green time

g_s = saturated green time

h = average spacing of vehicles in a queue on a traffic-signal approach

L = total lost time per cycle

n_r = number of right-turning vehicles

N = average queue at the beginning of the green period

N_s = initial queue at the beginning of a saturated green period

N_u = initial queue at the beginning of an unsaturated green period

q = flow on a traffic-signal approach

q_0 = flow on an opposing traffic-signal approach

q_1 = flow on the major road

371

q_2 = flow on the minor road

$q_{1(i)}$ = flow in the ith time interval of a predicted platoon at the stop line

$q_{2(i)}$ = derived flow in the ith time interval of a predicted platoon at a point along the road

r = turning radius

r_e or r = effective red period (cycle time − effective green time)

s = saturation flow

v = free running speed of vehicles on a traffic-signal approach

w = width of approach

x = degree of saturation; that is, the ratio of actual flow to the maximum flow

y = ratio of flow to saturation flow = $\dfrac{q}{\lambda s}$

y_{max} = maximum value of y

Y = sum of the maximum y values for all phases comprising the cycle

Y_{pract} = 90 per cent of Y

α = average gap accepted in the opposing flow by right-turning vehicles, or the mean lag or gap in the major-road stream that is accepted by minor-road drivers

λ = ratio of effective green time to cycle time

ndex

A' weighting curve 230
accidents, at priority intersections 150
    due to change of junction control 182
air pollution, due to carbon monoxide 246
    due to lead compounds 247
    due to oxides of nitrogen 247
    due to road traffic 245
    due to smoke 247
    due to unburnt fuel 246
area traffic control, Combination method of 366-7
    Equisat method of 367
    Flexiprog method of 367
    Glasgow scheme of 365
    Plident method of 367
    Toronto scheme of 363
    West London scheme of 364
assignment, all or nothing 65
    by diversion curves 65
    capacity restrained 67
    multipath proportional 68
    public transport 68

Bedford-Kempston transportation survey 9-11

capacity see Highway capacity, Intersection capacity
Category Analysis 28, 33
conflict points 273
congestion 250
counting distribution 108

decibel, definition of 229
demand curve 256-9
differential fuel taxes 252
dwelling insulation 238

Electronic metering 252-3
Evaluation of transportation surveys 71
    effect of running costs on 76
    use of first year rate of return in 74
    use of present value method in 74, 86
    use of representative flow in 75
    use of speed/flow relationships in 77-80

Fluid flow analogy of traffic flow 129

Glasgow area traffic control scheme 365

Headway distribution, congested 111-13
    counting 108
    double exponential 115, 172
    Erlang 112, 113
    negative exponential 108, 172
    Pearson 112, 113
    Poisson 108
    space 107
    time 107
    travelling queue 114
Highway capacity, effect of alignment on 101
    effect of gradient on 101-4
    effect of lane width on 100
    effect of lateral clearance on 100
    level of service 96-7
    maximum hourly volumes of 98, 99
    rural 98
    service volumes for 97
    speed/capacity relationship for 76-8, 125-32
    urban 99
    volume/capacity ratio for 96
Home interview survey 14-16

Intersections, accidents at priority 150
    accidents due to change of control at 182
    capacity of rural 165
    capacity of urban 166
    cloverleaf 223
    critical lag at 160
    delay at priority 163-5
    delay at traffic-signal-controlled 324
    diamond 221

Intersections—*cont.*
    direct connection 224
    effect of shape 212-13
    grade separated roundabout 222
    lag and gap acceptance at 156-9, 175
    new forms of single level 211-16
    partial bridged rotary 221
    partial cloverleaf 222
    simulation of traffic flow at 171
    three level 224
    trumpet 220
    weaving action at 183-7, 189-200

Jam density 126

Level of service 95
London Noise Survey 231
London Transportation Study 8, 32, 45, 59, 73
Los Angeles Regional Transportation Survey 8, 31

Marginal trip cost 259
Maximum hourly highway volumes 95
Methodology of transportation surveys 3, 4
Modal split 53-61
Moskowitz diversion curve 66, 67
Moving car observer method 145-8

Noise, effect of commercial vehicle content 234
    effect of gradient on traffic 235
    subjective reaction to 231
Noise due to vehicles 239
Noise levels 231
    measurement of 229
    prediction of 233, 234
Noise pollution level 232
Noise screening, by long barriers 238, 239
    by symmetrical short barriers 240
    by unsymmetrical short barriers 240

Origin-destination surveys 18, 19, 20

Parking garages, service times at 207
Postcard surveys 20
Price elasticity 256
Private cost 258

Quality of flow 193
Queueing theory, application to traffic flow 201–10

Representative flow 75
Restraint on vehicular traffic 250
Road pricing 251
        by differential fuel taxes 252
        by electronic metering 252, 253
        by supplementary licences 252
        economic benefits of 253, 259–61
Roundabouts, compared with traffic signals 182
        design of 182–6
        small island 211–16

Screenline 9
Simulation of traffic flow 171–80
Speed, analysis of studies of 137
        determination by moving car observer method 145–8
        flow and density relationships 125
        flow relationships 76–8, 125–32
        measurement of 136
        space mean 135
        spot 135
        time mean 135

Toronto area traffic control scheme 363
Traffic noise index 231
Traffic signals, capacity of 311
        combination method of linking of 356
        comparison with roundabouts 182, 211
        co-ordination of 342
        delay at 324
        early cut-off at 302
        effective green time at 291
        intergreen period at 278
        late start facilities at 302
        linking of 354
        lost time at 292
        optimum cycle time at 294, 329–33
        passenger car equivalent at 287
        phasing of 275
        platoon dispersion at 349, 352
        queue lengths at 336–9

Traffic signals—*cont.*
   right-turning vehicles at 306
   saturation flow at 282
   time and distance diagrams for 344
   timing diagrams for 299
   ultimate capacities of 311
   varying saturation flow at 319
   vehicle actuated facilities at 280
   warrants for 271

Transportation surveys, bus passenger 23
   commercial vehicle 4, 22
   evaluation of 71-91
   external cordon of 4, 18, 19
   home interview 4, 14
   methodology of 3-6
   origin–destination 18-20
   postcard 20
   screenline 9
   study area of 8
   taxi 4
   traffic zones of 4, 8
   train passenger 23
Travel costs, variations with speed 79
Travelling queue headway distribution 114
Trip attraction 42
Trip benefits 253, 259
Trip distribution, average factor method 35, 36
   competing opportunities model 46
   constant factor method 35
   deterrence function 42
   doubly constrained model 43
   Fratar method 35, 36
   Furness method 35, 38
   gravity models 41, 42, 43, 44
   intervening opportunities model 46
   production constrained model 42
Trip generation 6, 25, 26, 42

Weaving, at intersections 183-7
   multiple 190, 196-200
   operating conditions in 193

Weaving—*cont.*
    out of realm of 195
    quality of flow during 193
West London area traffic control scheme 364
West Yorkshire Transportation Study 8
Wilson Committee recommendations 232

Zonal least squares regression analysis 27